D1037384

Captivity and Behavior

Primates in Breeding Colonies, Laboratories, and Zoos

Captivity and Behavior

Primates in Breeding Colonies, Laboratories, and Zoos

Edited by

J. Erwin, Ph.D.
Humboldt State University
Arcata, California

Terry L. Maple, Ph.D.
Georgia Institute of Technology
Atlanta, Georgia

G. Mitchell, Ph.D.
University of California
Davis, California

Van Nostrand Reinhold Primate Behavior and Development Series

VNR VAN NOSTRAND REINHOLD COMPANY
NEW YORK CINCINNATI ATLANTA DALLAS SAN FRANCISCO
LONDON TORONTO MELBOURNE

Van Nostrand Reinhold Company Regional Offices:
New York Cincinnati Atlanta Dallas San Francisco

Van Nostrand Reinhold Company International Offices:
London Toronto Melbourne

Library of Congress Catalog Card Number: 78-26178
ISBN: 0-442-22329-3

Manufactured in the United States of America

Published by Van Nostrand Reinhold Company
135 West 50th Street, New York, N. Y. 10020

Published simultaneously in Canada by Van Nostrand Reinhold Ltd.

15 14 13 12 11 10 9 8 7 6 5 4 3 2 1

Library of Congress Cataloging in Publication Data

Main entry under title:

Captivity and behavior.

Includes index.
1. Primates–behavior. 2. Wild animals,
Captive–Behavior. I. Erwin, Joseph.
II. Maple, Terry. III. Mitchell, Gary D.
QL737.P9C29 599'.8'05 78-26178
ISBN 0-442-22329-3

Dedication

This volume is dedicated to several friends of ours without whom our research would have been impossible: Mr. Harvey H. Sveldt, Ms. Molly Jones, Walter, Julie, Ali, Sungei, Lunak, Sgt. Pepper, and Lucky Pierre; and goodnight, 71301, wherever you are.

Preface

We have always been concerned about the physical and social surroundings in which the primates we have studied have been housed. In fact, from the beginning of our research we have recognized the powerful role of environmental influences on behavior and development. We have provided information on a number of factors reputed to affect human development and behavior, especially those factors suspected of producing such adverse consequences as mental retardation, emotional disturbances, or personality disorders.

As we pursued the principles and processes underlying development of behavioral and social adequacy, we learned more about our kinship with the nonhuman primates, and we came to appreciate interactions between primates and environments and between primates as environments. From our research in laboratories, zoos, and breeding colonies, we have selected information we believe will lead others to share our concern over the impact of captivity on behavior. Most of the contributions included in this volume are drawn from a symposium on environmental influences on behavior of nonhuman primates that was presented at the 1977 meeting of the Western Psychological Association in Seattle, Washington. This symposium was arranged by the senior editor of this volume while he was at the University of Washington.

This volume is intended for the use of colony managers, animal scientists, veterinarians, zookeepers, scientists, and all others who are concerned with the design, evaluation, or maintenance of animals in captive environments. We trust that the volume will serve as a source of ideas as well as a fund of advice and information.

We are deeply indebted to the many friends, students, and colleagues who encouraged us and participated in gathering, analyzing, and interpreting the data from our research projects. Many of these helpful people are acknowledged specifically in the footnotes at the beginning of each chapter, but a few of those individuals who were especially influential in inspiring this work deserve special credit. Although they bare no responsibility, directly, for the contents of this book, we express our gratitude to the following people: Robert Sommer, Harry F. Harlow, William A. Mason, Gene P. Sackett, Thelma Rowell, Donald G. Lindburg, Orville A. Smith, Loring Chapman, Roy Henrickson, Gerald A. Blakely, and Mort Silberman. The pivotal direct contribution was made by Nancy Erwin, who gathered data at several stages of the research; she prepared most of the manuscript and assisted unselfishly in more ways than can be told.

Our appreciation to Alberta Gordon and Ashak M. Rawji for their patience, support, and constant attention to the details involved in the production of this volume.

J. ERWIN

Contents

1
Strangers in a Strange Land: Abnormal Behaviors or Abnormal Environments?*

J. Erwin

Department of Psychology
Humboldt State University
Arcata, California

and

R. Deni

Department of Psychology
Rider College
Lawrenceville, New Jersey

> In its natural situation, each species occupies a reasonably well-defined ecological niche wherein individuals tolerate or readily adapt to most variations in the environment. The species can compensate for minor environmental fluctuations physiologically and/or behaviourally; major environmental changes demand major behavioural changes.
>
> M. W. Schein and E. S. E. Hafez, 1969, P. 65

INTRODUCTION

We have frequently heard the assertion that nothing of importance can be learned from the study of primates in captivity, especially with regard to their behavior. It is contended that the behavior of captive primates (and other animals) is distorted beyond recognition by rearing and/or maintenance in unnatural environments. Despite this criticism, there are many primates and other animals on display in zoological and animal parks, scientists continue to use large numbers of animals in laboratories for biomedical and behavioral research, and

*Preparation of this manuscript and some research referred to in it was partially supported by USPHS/NIH grants HD00973-12 and HD04510 to IMRID and the Kennedy Center, George Peabody College for Teachers.

the public is consistently attracted to and fascinated by animal acts from wrestling bears and dancing dogs to chimpanzees doing stunts on motorcycles. Why are these practices perpetuated if nothing of value is learned from them? Obviously, there is disagreement over the importance of answering various questions, and there are certainly differing viewpoints regarding the choice of appropriate questions.

It is quite a different matter to ask "How do chimpanzees communicate in their natural setting?" than to ask "Are chimpanzees capable of learning to use language for communication?" or "Of what adaptive value has the development of gestural communication been for chimpanzees?" Some questions are best asked in the context in which adaptation occurred; others require environmental control and manipulation, or at least unhindered visual access, not often possible in free-ranging situations. One approach emphasizes description of species-typical behavior and attempts to establish ultimate causation; another is more concerned about immediate or proximate causal influences. Study of the effects of captivity on behavior requires both of these approaches and a variety of comparisons. In this volume, we offer comparisons of the behavior of members of various conspecific dyads under similar physical conditions (Chapter 3), cross-species comparisons under similar circumstances (Chapters 4 and 7), comparisons of spatial and social influences (Chapters 5 and 6), and comparison of reactions of various primates to deprived and enriched captive conditions (Chapters 8 and 9). A history of maintenance and use of primates in captivity is also included (Chapter 2).

Much of the material included in this volume is oriented toward behavioral potentials rather than behavioral typicalities as expressed in natural habitats. Ideally, evaluation of effects of captivity on behavior would be accomplished by direct comparison of species-typical behavior in the wild with the form and frequency of the same behaviors in various captive settings, along with documentation of those unusual or maladaptive behaviors that might occur in either circumstance. Direct comparisons have not usually been available (but see Chapter 7); consequently, comparisons between and among captive settings offer the majority of information available. Clearly, those factors that are operational in captive environments are *potentially* operational in the natural setting, and the definition of factors involved in proximate causation can generate valuable questions regarding ultimate causation.

Questions regarding behavioral potential or proximate causation are usually investigated in captivity in situations allowing controlled environments. They deal primarily with behavioral and physiological mechanisms. While studies of behavioral potential may discover that chimpanzees can, indeed, learn to use language, we are immediately prompted to ask "by what evolutionary mechanism could the capacity to learn language precede the invention of language?" *The potential to perform any act must precede the actual performance of that act.* Thus, the study of behavioral potentials and the mechanisms by which the processes operate are of primary importance to evolutionary biology.

The study of behavioral potential (see Kuo, 1967) is dual. We are asking about the capabilities of *organisms*, e.g., for language learning by chimpanzees, *and* we are concerned about *processes*, e.g., the process of learning language. We are interested in the chimpanzee for its own sake *and* for what it can teach us about processes.

The consequences of rearing primates in captivity are sometimes distasteful, yet the systematic laboratory studies have helped to provide methods of preventing and treating behavioral pathologies associated with deprivation and confinement in human, as well as nonhuman, primates (Baumeister and Forehand, 1973; Berkson, 1967). Abnormal behaviors exhibited by monkeys and apes in zoos and laboratories in many ways resemble those behaviors commonly seen in institutionalized humans (cf. Davenport and Menzel, 1963; Erwin et al., 1973; Mitchell, 1970; Sackett, 1968).

It is the purpose of this chapter to review some of the information on abnormal behavior of primates and the environmental factors associated with them. First, we will discuss some definitions of abnormal behavior; next we will describe a number of environmental factors that may influence behavior; and finally, we will cite examples of the role of environmental influences in the development or prevention of maladaptive behavior among captive primates.

DEFINITIONS OF ABNORMAL BEHAVIOR

Sackett (1968) pointed out the relativistic nature of the terms abnormal and normal. Clearly the definition of abnormal is dependent upon the definition of normal, and for studies conducted in captivity, Sackett suggested the term "abnormal" could be used appropriately with "... a particular control condition as a reference–and not necessarily a 'normal control' group–... it seems reasonable to collect normative data on animals reared in any condition and to compare the quality and quantity of their behavior with that of animals reared under different conditions." While we have no doubt that such comparisons are appropriate, and have employed that method ourselves, it seems to us that the range of conditions should include the natural habitat or close approximations thereof wherever possible. We realize that a similar requirement for definition of normal human behavior would require field studies of primitive human tribes in natural settings, and we believe that such studies are essential. At the same time, we cannot ignore the behavioral potential of humans in artificial surroundings such as cities.

It is certainly possible and appropriate to gather normative data for each of many environmental conditions, as Sackett (1968) has suggested. A hierarchy of frequencies, durations, or average durations of behaviors could be constructed for each setting. The most prevalent behaviors would be considered normal. Doubtless, we would see a shift from adaptive behavior in natural settings to maladaptive behavior in restrictive situations, i.e., we would find that behaviors that were "normal" in the natural setting became "abnormal" in the restrictive setting, and that behaviors that were uncommon ("abnormal") in the relatively natural situations would be "normal" (prevalent) in the restricted setting.

Now that we have made that point perfectly obscure, there is a need to restrict the definitions of normal and abnormal behavior. Surely we do not simply mean that a behavior is adaptive, healthy, or desirable when we use the term "normal," but neither do we mean *simply* that the behavior is usual, frequent, or common, regardless of the situation, even though that is the most appropriate literal use of the term normal. When we refer to "abnormal" behaviors we do not merely mean uncommon; we probably are referring to aberrant behaviors or behavioral pathologies, such as stereotyped movements, self-biting, bizarre postures, hyperaggressivity, etc. We think of these as pathological behaviors because they occur under restricted conditions rather than in natural habitats, and because psychotic, emotionally disturbed, or retarded humans in institutions display these behaviors. In this chapter, we will use "abnormal" to denote behavioral pathologies rather than to refer to infrequently occurring behaviors (except in the sense that these "abnormal" behaviors occur infrequently under free-ranging conditions).

Another essential element for determining behavioral adequacy suggested by Sackett (1968) is "ecological validity." He suggested that the test environment should be appropriate for the usual housing conditions or the rearing condition of the individual being tested. Laboratory animals are typically less reactive to laboratory testing situations than are wild-reared animals. This is probably also true of more permanent enclosures such as those in zoos; animals reared in the situation are likely to be better adapted than those introduced from other situations, either less or more restrictive. Thus, for many laboratory tests, the laboratory-reared subject is most appropriate.

Some abnormalities, as we will describe below, are apparently transient and are primarily situational; that is, they occur in response to the environment rather than to persistent individual deficits such as those associated with deprived rearing conditions. Berkson (1967) has distinguished between "cage stereotypies" and "deprivation acts" in essentially that manner, i.e., cage stereotypies are responses to restrictive conditions, and deprivation acts are long-term developmental disorders associated with rearing in restricted environments. Mitchell (1970) has listed several rearing experience factors that contribute to the development of behavioral pathologies: 1) age of the animal at the time of exposure to the treatment; 2) the age of the animal at the time of posttreatment testing; 3) the duration or quantity of the treatment experience; 4) the type or quality of the treatment experience; 5) the relation of the treatment to the genetic background of the animal; and 6) the sex of the animal. Posttreatment testing factors can also influence the expression of abnormal behaviors, and Mitchell (1970) has suggested the following factors: 1) differences in the type of general nonsocial environment employed in testing (e.g., novel versus familiar; large versus small); 2) differences in the length of test sessions (e.g., the longer the test sessions, the more likely it is that affiliative behaviors will be seen); 3) the type of scoring system used; 4) interobserver reliability; 5) type of data used (frequency, latency, duration, etc.); 6) the number of social partners provided; 7) the species of social partners provided; 8) the age and sex of social partners; and 9) familiarity of social partners. Specific studies describing demonstration of effects of these factors are included in this volume.

DESCRIPTIONS OF ABNORMAL BEHAVIORS

There are two basic classes of behavioral pathologies. They are qualitative abnormalities and quantitative abnormalities, relative to behavior in natural settings. Qualitative abnormalities are those that occur in captivity but not in natural settings; quantitative behavioral abnormalities are those that occur more or less often in captivity than in natural settings. Some qualitative abnormalities seem to be related to normally occurring behaviors, but are distorted and are clearly not adapted to the captive situation, or the function in the captive environment differs from that of the natural environment. Berkson (1968) has referred to these behaviors in captivity as "homologous" with normal, adaptive behaviors, apparently because they are based on the same organization of motor patterns and are often similarly motivated.

Qualitatively Abnormal Behaviors

Bizarre Postures. There are a number of postural disturbances that appear in confined primates. Some seem to serve self-stimulating functions for animals in restricted environments. These disturbances are especially common in (but are not restricted to) animals reared in isolation.

Floating limb. Isolate-reared rhesus monkeys frequently display an odd pattern of movement of a limb. Usually while sitting quietly, one limb will begin slow, upward movement. This apparently goes unnoticed, at first, by the animal itself. Then the animal seems to become aware of the floating member via peripheral vision; after visual tracking for a short time, the floating limb is often attacked viciously.

Self-biting. Several patterns of self-biting have been observed; most commonly one hand is bitten in a very stereotyped fashion, but arms, legs (see Fig. 1-1) and torso are also targets of self-biting. Species and age differences in this behavior have been documented. While self-biting is rather common among isolate-reared rhesus, it is by no means restricted to animals with impoverished early experience

Fig. 1-1. Self-biting by a laboratory-born adolescent rhesus male. This animal was reared with access exclusively to his mother for the first 8 months of life; shortly after separation from his mother he was matched for age and sex with a peer, with which he was paired until the end of his second year.

(Erwin et al., 1973); self-directed aggression is frequently elicited by frustration (Gluck and Sackett, 1974; Tinkelpaugh, 1928).

Self-clasping and self-grasping. Self-clasping (embracing) and self-grasping (use of hands or feet to hold onto a part of the body) are especially common in isolate-reared rhesus monkeys, but these patterns are not exclusively confined to animals deprived of social contact early in life (Fig. 1-2). These behaviors are, however, probably homologous with mother-directed behavior and eventually sexual clasping and grasping. The etiology for mother-deprived animals

Fig. 1-2. Self-clasping by another rhesus male reared under the same conditions as the animal shown in Fig. 1-1.

seems reasonably straightforward, but it is difficult to explain the genesis of this pattern in subjects receiving maternal and peer-sequential social experience. Perhaps during the mother-only period of rearing the responses become habitual and later become self-directed during periods of isolation or lowered stimulation.

Saluting. Eye poking, probably a form of self-stimulation, is quite common among socially deprived rhesus monkeys, but has also been seen in mother-peer sequentially reared rhesus (Fig. 1-3), and in group-reared crabeater macaques. The behavior is typically asymetrical in the sense that it is usually done only on one side and always the same side, although we have observed simultaneous, bilateral saluting in nonisolate-reared rhesus in the Peabody colony.

Stereotyped Motor Acts. The most common type of abnormal behavior associated with captivity is stereotyped movement. It is not very unusual to find some forms of stereotypy in wild-reared animals.

Stereotyped pacing. A major form of activity for captive primates is stereotyped pacing. The patterns of pacing are idiosyncratic but are very well defined within the individual. They may involve movement back and forth along one wall of a cage (often along the front of the cage), circular movement around the cage, walking forward or backward, a figure-eight pattern, or any combination of these. Stereotyped pacing patterns are often integrated with the other stereotyped acts described below. These patterns are not confined to animals reared in captivity, but virtually all captive-reared macaques confined in small cages early in life express this behavioral pattern.

Head tossing or weaving. As a part of stereotyped pacing patterns, it is common to observe head tossing. The stereotyped head-tossing pattern often occurs at the end of a repetition of a stereotyped pacing pattern. Weaving, movement of the head and shoulders from side to side, is also a common occurrence in captive rhesus.

Bouncing in place. Young rhesus monkeys sometimes develop habitual jumping up and down, usually on all fours, when they are

Fig. 1-3. "Salute" posture exhibited by another nonisolate-reared rhesus monkey called Sgt. Pepper.

confined alone in cages. The behavior appears to develop rapidly and probably serves a self-stimulating function. There is considerable similarity of pattern in this behavior across individuals.

Somersaulting. Repetitive somersaulting is another stereotyped locomotory behavior that is very common in laboratory-reared rhesus. This pattern is usually very similar across individuals. Somersaulting apparently provides some badly needed exercise for laboratory animals, and probably involves vestibular self-stimulation.

Rocking. Rocking is often associated with self-clasping and occurs primarily in isolate-reared primates. Recent work by Mason and Berkson (1975) has demonstrated that rearing on a mechanical surrogate mother that moves (swings) eliminates the development of rocking stereotypies. Rocking apparently also involves vestibular self-stimulation. Some forms may, however, be homologous with sexual thrusting and may be maintained through sexual self-stimulation.

Appetitive Disorders. Some appetitive disorders are probably qualitatively distinct from normal behaviors under natural conditions. Others (described in the next section) are quantitative problems.

Coprophagia. It is rather common for nonhuman primates in captivity (and some institutionalized human primates) to habitually ingest fecal material, not necessarily their own. This strikes us as a disgusting habit and one that should be eliminated. It may stem from dietary deficiencies or may be gustatory self-stimulation.

Paint eating. Virtually any painted surface is likely to be destroyed by nonhuman primates due to their habit of eating the paint. Lead-free paint must be used to avoid lead poisoning. Other surface materials are also frequently ingested. Acrylic plastic is usually scratched at and gnawed at by rhesus macaques in laboratory settings. The residue collected is ingested, and the scraping sound made in the process is extremely irritating to the experimenter. When possible, tempered plate glass is preferable to acrylic plastic in cage or display construction for primates. Extreme care is essential regarding materials placed in captive primate enclosures, because virtually anything is likely to be ingested (e.g., broken glass, mortar, concrete).

Sexual Disorders. There are many sexual disorders associated with captivity in primates, as suggested by the difficulty of initiating breeding programs and perpetuating breeding across generations born in captivity. We will list a few examples here.

Inappropriate orientation. Both male and female rhesus macaques reared under restrictive conditions, particularly social isolation, often fail to establish appropriate sexual posturing for copulation. The species-typical pattern for rhesus is rather stereotyped, involving a presenting posture by the female, mounting with ankle clasps by the male, insertion of the penis (often with assistance from the female), thrusting by the male, reaching back and grasping the thigh of the male by the female, and facial orientation with lipsmacking. Not all these elements are essential to successful copulation, but animals deprived of early social contact often attempt to mount the head or the side of their companion rather than establishing appropriate orientation. Many isolate-reared animals appear to be highly sexually motivated despite their inability to perform appropriately.

Homosexual behavior. There has been considerable confusion of sexual and dominance behaviors of primates. It is not uncommon for macaques to mount members of the same sex. It has been hypothesized that this action aids in establishment and reaffirmation of dominance status within social groups. It is clear, however, that some of this behavior is sexually motivated, because there is sexual arousal and thrusting, and, in some cases, anal penetration of one male by another (see Chapter 6). It is not clear whether exclusively homosexual relationships form under free-ranging conditions; the laboratory rhesus males were not exclusively homosexual, although they were found to prefer each other to an unfamiliar but highly receptive female. This condition may not be pathological, but it is abnormal in the sense of not being species typical.

Sexual dysfunction. A common problem in captivity is depression with associated loss of sexual motivation or function. This probably results from general apathy in nonstimulating environments in which behavioral activity is unrewarding, and it may also be associated with lowered hormonal levels such as those sometimes found in depressed humans. Primates that ordinarily exhibit seasonal breeding patterns may become totally unproductive in captive conditions that simulate the low period of sexual activity in the natural setting.

Autoerotic stimulation. In some captive primates, especially those reared under conditions of social deprivation, sexual activity is primarily self-directed. It is especially common for male infants to develop a habit of penis sucking. While this may at first be homologous with nursing behavior, it is apparently *maintained* as sexual self-stimulation. Other types of autoerotic stimulation are probably not qualitatively distinct from naturally occurring patterns.

Quantitatively Abnormal Behaviors

Most behavioral patterns probably occur at different rates in captivity than under free-ranging conditions, even if their form is not distorted to the point that constitutes a qualitatively distinct pattern. The degree to which a quantitative difference must deviate from species typical before it is recognized as abnormal, pathological, or maladaptive varies according to the behavioral pattern and the risks associated with it. Examples of some quantitatively abnormal behavioral patterns are described below.

Activity Patterns. A major problem among captive primates is apathy and depression. Inactivity can often be remedied by provision of contingencies that encourage activity (see Chapter 8). It is possible to regulate activity to desirable levels using such systems. Hyperactivity is usually expressed as stereotyped behavior as described in the previous section of this chapter. Temporal distortion in behavioral patterns can result from artificially extreme light-dark cycles, particularly from patterns involving constant bright light.

Appetitive Disorders. Quantitative disorders of appetite involve hyperphagia (uncontrollable eating), hypophagia (insufficient eating or "finickiness"), and polydipsia (frequent drinking). Polydipsia is common among isolate-reared rhesus monkeys by contrast with wild-born conspecifics, but the mechanism by which polydipsia develops has not been specified. There is some evidence from work on schedule-induced polydipsia that frequent drinking arises during periods when the environment is unresponsive to operant behavior. Hyperphagia may result from unresponsive or unchanging environments, or nutritional disturbances, and hypophagia may be associated with depression or lack of exercise. Refusal to eat or drink is often associated with illness, environmental changes, or special characteristics of the food or water.

Agonistic Disorders. Hyperaggressivity is common among isolate-reared primates, probably due to the lack of development of social attachments with conspecifics. Wild-born animals also display more aggression in captivity than in free-ranging situations, but serious aggression does occur in natural settings. Some influences of social and spatial environments on expression of aggressive behavior are discussed in this volume (Chapter 6).

POTENTIAL ENVIRONMENTAL INFLUENCES

Each dimension of the environment has potential for influencing the behavior and even the survival of all organisms existing within the environment. There is a range along each dimension within which an organism can adapt and survive, but for any characteristic of the environment, too much or too little of that factor will be lethal. All organisms that exist are alive because their ancestors survived through a range of varying circumstances and endowed them with genetic potential for development and adaptation within a particular range of environmental constraints.

Under natural conditions, organisms can be observed at variously effective dynamic states of adaptation approximating equilibrium. Organisms must live in virtual equilibrium with an environment which consists of many variable factors, including abiotic and biotic factors (Vernberg and Vernberg, 1970). The abiotic component of the environment consists of such chemical and physical factors as oxygen, temperature, pressure, pH, etc., and the biotic component includes all interorganismic relationships, such as agonistic behavior, social attachments, predator-prey relationships, symbiosis, etc. In the natural habitat many generations of adaptation and survival usually enable survival of a sufficient number to propagate the species.

Abiotic Factors

Some of the most notable abiotic factors will be mentioned here. In creating captive environments it is essential, of course, that we control each factor within a range compatible to the captive population. Additional temporal variation should also be appropriately regulated.

Temperature. The temperature of an environment is, of course, a measure of heat energy, and the amount of heat energy in an environment is critical for some physicochemical reactions to occur. Within natural settings temperature varies according to an annual cycle (seasonal variation) and a daily cycle (as well as other, longer, but less consistent cycles). Primates and other mammals have the capacity to regulate their body temperatures physiologically and behaviorally within reasonable constraints, but extreme deviations can be dangerous. Sustained high temperatures are stressful and may result in dehydration; sustained low temperatures are also stressful and may lower resistance to several disease processes, or may even result in freezing. Rapid temperature changes may be especially stressful. Although data are not yet complete, it seems likely that, in some seasonally breeding primates, temperature may be a critical factor in the reproductive process. There may be temperature thresholds that must be maintained on a daily or seasonal basis to ensure gonadal activity or mating behavior or both. A failure to provide thermal patterning typical of the natural environment might block breeding in captivity, but there is a need for additional study with most primate species on the influences of temperature on behavior, particularly within the ranges usually encountered in natural or captive settings (see Bernstein, 1972). It is clear that a broad range of temperature tolerance occurs in primates, ranging from the Japanese snow monkeys to exclusively tropical species.

Illumination. Most heat energy is supplied by radiation within the range of humanly visible wavelengths (the range also visible to most other primates). There are few nocturnal species of primates; all except one are prosimians (the exception is the New World monkey, *Aotes*, the owl monkey). Lighting is probably influential on many kinds of behavior, including reproductive behavior. This is particularly probable in seasonal breeders, and the length of the light portion of the light-dark cycle may cue seasonal breeding in some species. Most births occur during the dark portion of the cycle in those diurnal primate species about which birth data have been gathered. It is possible to change the timing of birth by altering artificial light cycles (Jensen and Bobbitt, 1967). Activity patterns, and other behaviors also, exhibit marked diurnal changes (Bernstein, 1972; Martenson et al., 1977), and it is also possible to alter these activity patterns by adjusting the light-dark cycle. One interesting application of this procedure involves monitoring visitation schedules of zoo visitors and activity cycles of animals exhibited. The light-dark cycles of animals exhibited in artificially lighted enclosures (such as the nocturnal primates) can be adjusted so that the animals are most active and interesting at the time of day when they will be visited most often. This procedure can, of course, also be used in the laboratory to provide optimally convenient observation times, particularly for daytime monitoring of parturition (Jensen and Bobbitt, 1967). The quality and quantity of illumination in captive settings, and particularly the variety and cyclicity of lighting, may exert powerful influences on the behavior of primates in captivity, but this problem has not yet received adequate attention by researchers.

Water. The prevalence of water in the captive environment can also influence the behavior of primates in several ways. In the natural setting, water is likely to be available only in specific locations such as rivers, springs, or lakes. The troop moves into watering places one or more times daily as a part of their foraging route. In captivity, water is usually available *ad libitum* from bottles or automatic watering systems. We (Erwin and Deni, unpublished data) have found considerable variation in water-drinking patterns of rhesus monkeys under laboratory conditions, with very low consumption at night and high consumption in midmorning (0800–1000) and midafternoon (1400–1600). As mentioned earlier, polydipsia has been reported for laboratory (isolate-reared) rhesus. This may be due, in part, to the continuous availability of water in the laboratory setting. Polydipsia is, however, not likely to be a serious disorder unless it is really extreme.

Another aspect of presence of water in the environment is humidity. There may be some effects of humidity on thermoregulatory behavior and some aspects of reproductive behavior, but these are not well documented. Seasonal rain in areas of origin may cue onset or offset of some reproductive behaviors, but the evidence on this matter is not yet clear. The degree of plasticity of such effects on mating and birth seasonality following transfer to new settings remains obscure.

Space. The amount of physical space available surely influences the behavior of primates and may even effect reproduction, although there is no evidence for the latter. Chapters 5 and 6 of this volume deal with some aspects of behavioral responses to variation in space. It should be noted, however, that the responses reported are primarily social in nature and are consequences of interaction of abiotic factors (amount of physical space) and biotic factors (presence of conspecifics). Decisions on amount of physical space required for adequate development must be empirically based, but, unfortunately, there is little information available other than that presented or cited in Chapters 5 and 6 of this volume. There is a tremendous need for research in this area, especially in the light of the need for domestic production of primates for biomedical use and the desire to improve laboratory and zoo environments. Environmental complexity is probably very important, particularly for rearing environments.

Other Factors. Many abiotic environmental factors other than those mentioned above are potential agents of influence on behavior, physiology, and development. Altitude and atmospheric pressure, for example, may influence development, including delay of sexual maturity of females. Care must be taken to avoid exposure of captive primates to polluted air or water, and, of course, contact with dangerous chemicals must be avoided. Captive primates will break and/or eat virtually anything they can get their hands on in captivity (unless one *wants* them to do so), so it is necessary to "monkey proof" structures and apparatus introduced into primate enclosures. We have seen glass lighting fixtures broken and eaten (with consequent intestinal perforations and peritonitis). In one of our studies, concrete pipes were partially eaten by macaques. While these responses may reflect boredom, they require that attempts to enrich environments involve making objects or materials available only if they are edible or sufficiently durable to prevent breakage and ingestion.

Biotic Factors

In addition to the factors listed above, the potential for influence from biotic factors in natural and captive environments must be recognized. Of course, biotic factors exert influence via the internal as well as the external environment.

Disease. A general class of biotic environmental influences on behavior and development is disease. Infections and infestations of various kinds may be first recognizable by behavioral correlates such as vomiting, lethargy, increased irritability (as indicated by aggressive behaviors), increased incidence of fear/pain-related behaviors, diarrhea, avoidance behaviors, crouching, etc. The use of behavioral/observational measures in diagnosis of primate illness is not yet widespread, but with further research over the next few years, the diagnostic value of quantified behavioral measures will probably become better recognized.

An important environmental consideration, with regard to disease factors, is the need for methods to restrict spreading of disease from one animal or group of animals to others. Animals brought into captivity are quarantined to reduce the likelihood that serious diseases will be introduced into captive colonies, but cases that do not become active during the quarantine period may result in exposure of many additional animals after introduction into the colony. Therefore, it is desirable to have some built-in barriers to transmission of disease. Gastroenteritic diseases are especially susceptible to transmission through contact of one animal with the feces of another. Thus, *Shigella flexneri* is likely to be transmitted to all the members of a social group if a carrier is introduced to a group. To the extent that groups intermingle or occupy the same space, the likelihood of transmission is increased.

At the University of Washington Primate Field Station, a series of housing strategies was tried before a suitable one was chosen. The facility was a prison renovated for use as a primate breeding colony. Macaques and baboons were maintained in harem groups with one group per cell. Small openings with sliding doors were fashioned to allow groups to be shuttled from one cell to the next for cleaning purposes. Thus, a room would be pressure washed, the group from the next room would be shuttled into it while their room was cleaned, etc., until all rooms were cleaned. During that period gastroenteritis was common in the colony, and it was believed by some that the cleaning procedure might have allowed too much contact among groups. The housing strategy was changed so that each group had access to two rooms and only two rooms. The cleaning procedure involved crowding all members of the group into one room of the two-room suite while the other room of the suite was cleaned, and likewise for cleaning of the other room. This isolation of groups from one another apparently reduced incidence of enteritis only slightly, but a possible contributor to continued incidence was that aggressive interactions (with consequent trauma and stress) were much more frequent when groups had access to two rooms than when they were confined to one room (see Chapter 6). The high incidence of trauma contributed to group instability, as did removal and reinstatement of experimental animals from the colony. Subsequently, a procedure was adopted in which each group was confined to a specific two-room suite (eliminating spatial overlap among groups, with increased risk of transmitting disease), but the group was given access to only one room of the suite at a time. Each day the group was moved to the clean room of their suite, and the dirty room was cleaned. Among other things, this allowed the room cleaned on one day an entire day to become completely dry, unlike the previous procedure. Food could be scattered around the clean, dry room to afford better access to all group members. Animals had to be moved only once each day, rather than twice, and the number of rooms cleaned each day was reduced by 50 percent, allowing a reduction in the amount of animal caretaker time devoted to cleaning.

Another procedure that minimized changes in group composition was also implemented at about the same time. Groups were ranked according to priority based on reproduction and freedom from trauma or disease. On the basis of this ranking system, animals were withdrawn for experimental use from low-priority groups; the highest priority groups were not subjected to unnecessary manipulation of social structure. The incidence of trauma and disease in the colony decreased substantially, and it is likely that the impact of other problems associated with stress were probably also reduced. This case is offered as an example of the complex relationship that usually exists among biotic factors in captive primate environments. A systems analysis approach to colony management and design of captive environments should include integration of information on housing conditions, maintenance procedures, potential disease agents, trauma rates, behaviors, and reproduction.

Food. Obviously, availability of appropriately nutritious food is necessary to support any population in captivity or under natural conditions, but it is important that the patterns of feeding in natural settings be imitated, if possible, in the captive setting (see Chapter 8, this volume). In wild primate groups, much of the day may be spent foraging. If animals are provisioned, several problems may ensue. A relatively short period of time each day is required for acquiring adequate food; consequently, animals may overeat and become obese, or they may occupy the noneating hours in some other fashion, such as fighting. Provisioning (or feeding, in zoos or breeding colonies) is often done in relatively limited space. This sometimes results in aggressive encounters over food that would not occur if feeding stations were more widely spaced. Feeding at one (or few) stations can also result in total exclusion of subordinate animals. It is desirable to spread feedings out in time, as well as space, to ensure occupation of more time in eating and access of all group members to adequate nutritional supply.

Presence of Conspecifics. Primates are sufficiently dependent at birth that they require considerable amounts of care during infancy. Thus, in nature, the primate life is social from the time of birth. The social patterns differ across species, as do the ways in which environments are used (see Fragaszy, Chapter 7, this volume). Much of this volume deals with patterns of social interaction in captivity. It is clear that the composition of groups and the disruption of group structure can have disastrous consequences. An even more devastating problem can result from the *absence* of conspecifics at various life stages. The final section of this chapter deals with effects of various patterns of rearing in captivity, including the effects of hand rearing and early weaning.

REARING IN CAPTIVITY

While biotic and abiotic environmental factors can be discussed separately, it is important to recognize that in reality all these factors interact during the development of each individual. The influences of these factors are exerted primarily when they are present in excessive or unusually diminished quantities. The dynamic relationship among various environmental influences is especially apparent as expressed during development, and many of the problems associated with captivity in primates are environmentally induced early in life.

Ruppenthal and Sackett (in press) recently reviewed some of the information of interest to those of us who are responsible for making primate husbandry decisions. They have pointed out that rearing conditions can not only influence survival, reproductive capabilities, and behavior, but also the suitability of primates as subjects for biomedical research. The recognition of potential rearing experience effects by medical researchers may increase the demand for specialized production of primates, but many problems remain to be solved to improve reproductive capacities of primates in breeding colonies and zoos.

Types of Rearing

The effects of specific rearing environments have been examined, primarily with regard to behavioral development, including reproductive behavior.

Total Social Isolation. Total social isolation environments include any situation in which an animal is removed from sensory access to conspecifics. A more extreme form of isolation can involve a total lack of interaction with humans or animals other than conspecifics. Total social isolation affects reproductive behavior, aggression, maternal behavior, exploration, play, learning, and eating and drinking habits. Isolate-rearing is discussed in more detail in Chapter 3 (this volume), but, in general, the effects are extremely detrimental. Many bizarre behaviors ensue, and adequate reproductive behavior seldom, if ever, occurs in males; females are deficient in sexual receptivity and are abusive to infants.

Partial Social Isolation. The standard procedure in some breeding colonies is removal of infants from their mothers at birth and rearing in separate cages. In these situations, individuals can often see, hear, and smell conspecifics but cannot touch them. This common rearing pattern is referred to as partial social isolation, and the consequences are nearly as bad as those of total social isolation. Bizarre behaviors are common, and appropriate sexual behavior is usually absent. Isolate-rearing should not be practiced with animals intended for subsequent breeding.

Mother-Only. Another common technique for primate breeding colonies and laboratories is maintenance of the mother and infant together in an individual cage until the infant is weaned (usually weaning is accomplished artificially by separating the mother and infant). Infants are weaned at various ages. Mother-only reared infants exhibit fewer abnormalities than do those that are isolate-reared. They seldom rock, but some other aberrant behaviors are fairly common. If mother-only experience is continued as the exclusive form of social contact, later adjustments to social situations may be poor (Ruppenthal and Sackett, in press). The surrogate mother-reared monkeys studied by Harlow (1958) exhibited grossly abnormal behaviors during juvenility and adulthood (Sackett and Ruppenthal, 1973); some aberrant behaviors, such as rocking, were eliminated by provision of a mobile surrogate (Mason and Berkson, 1975) rather than a stationary one.

Peer-Only. Rearing infant monkeys without maternal contact, but with access to peers instead, has been attempted (Chamove, 1966) and has been found to depress the appearance of typical play behavior, apparently due to the development of persistent clinging ("together-together" syndrome). The clinging pattern occurs also in peer groups of infants. Although the presence of peers is important (cf., Kuyk et al., 1977), it appears that peer-only rearing is insufficient for optimal social and sexual adjustment in adulthood.

Mothers and Peers. In some laboratories infants are reared first with their mothers (in individual cages). At some point, they are separated (e.g., six or eight months of age) and are placed in cages containing one or more peers. Sociosexual behavior is fairly adequate in rhesus monkeys with this kind of background (Erwin and Mitchell, 1975), and maternal behavior is adequate in individual cages. There are two basic patterns of mother-peer rearing. One is *mother-peer sequential* (as described above); the other is *mother-peer simultaneous.* In the latter condition, infants have access to their mothers and other infants housed in groups. In some laboratories cages are arranged to allow infants access to their mothers and to peers and the peers' mothers, but the adult females do not have contact with each other. As Mitchell (1970) has suggested, mother-peer rearing is nearer optimal than either mother-only or peer-only rearing. Animals reared in this manner commonly exhibit some bizarre or stereotyped behaviors (Erwin et al., 1973), although fewer types of aberrant behavior occur than in isolate-reared rhesus monkeys, and those abnormal behaviors that do occur are exhibited less frequently.

The Nuclear Family. As Mitchell (1969) suggested (also see Chapter 3, this volume), adult males also play some role in the socialization of infant primates, as do siblings, adolescents, and adult females other than the mother. Suomi (1977) has studied the nuclear family rearing situation in detail. Nuclear family groups with playroom areas between family cages (Harlow, 1971) have allowed study of the development of rhesus monkeys in more complex settings than those usually available in the laboratory (much more like the zoo setting), and it seems likely that animals reared in this situation will prove to be better socially and sexually adjusted than most other laboratory-reared monkeys.

Harem Groups. The most promising type of captive group for breeding purposes is probably the harem group (for appropriate species) as described in Chapter 6 (this volume). In this setting, one adult male and several females (with their offspring) are housed in the same enclosure. Paternity is known because only one male has access to all the females. Infants are exposed to a complex of appropriate sociosexual stimuli which should prepare them well for further group living and reproduction.

Large Heterogeneous Groups. Some of the captive breeding colonies have formed groups in large corrals or on islands. At the Delta Regional Primate Research Center, for example, groups were started by introducing five adult males and forty-five adult females into each large corral. After such groups become reasonably stable, they should provide excellent rearing environments and high-quality reproductively capable offspring.

Other Factors. There are a number of other management practices that influence the behavior and development of captive primates. Some of these are described briefly here.

Effects of cage size. Draper and Bernstein (1963) tested the effects of cage size on stereotyped locomotion in wild-reared rhesus monkeys. They found that more stereotyped locomotion occurred in small than in large cages. More recently, Paulk et al. (1977) tested laboratory-reared and wild-reared rhesus in large and small cages. They found stereotyped locomotion in all adults that had been laboratory-reared and in one-third of the wild-born adults in their study. This result agrees well with our findings from the Peabody rhesus colony in which we found that all laboratory-reared animals exhibited stereotyped motor acts, while a substantial number of wild-born rhesus that had been housed in the laboratory for several years did not perform any stereotyped locomotion. Paulk et al. (1977) found that stereotyped locomotion increased in the smaller cage relative to the larger one, while nonstereotyped locomotion decreased. Nonstereotyped locomotion also decreased in the individuals that displayed no stereotyped locomotion. Thus, it appeared that the effect on normal locomotion was opposite that for stereotyped locomotion. Cage size influenced both the quantity and the quality of activity in rhesus monkeys. Other aberrant behaviors were not affected by cage size.

The studies reported above substantiate short-term effects of cage size and suggest a developmental component in the etiology of stereotyped motor acts. Castell and Wilson (1971) reared *Macaca nemestrina* infants with their mothers in large or small cages and compared these animals with others living in a group situation. The separation process of mother-infant dyads differed, with group-reared infants remaining near their mothers more during early development than did cage-reared infants. The development of independence occurred earlier for group-reared than cage-reared *S*s. Infants in small cages were punished more than were infants in the other situations. Jensen et al. (1971) found that the effects of rearing in a small impoverished environment persisted into adolescence. Riesen et al. (1977) and Struble et al. (1977) reared infant stumptail monkeys in four conditions of crowding and complexity ranging from a 30-cm^3 plexiglas cube to a social group. Motor-deprived (small cage) animals were highly emotional relative to controls but were less active or manipulative. Males were more affected than were females.

Multiple separations. Repetitive separation of infants from their mothers can have lasting consequences on the behavior of infants as demonstrated for rhesus monkeys by Griffin (1966), Mitchell et al. (1967), and Ruppenthal and Sackett (in press). When tested in social situations following repeated separations from their mothers, infants were subordinate to control animals that were not repeatedly separated and to other infants that had been repeatedly fostered to adult females other than their natural mothers. One year after the final separation, animals repeatedly separated from their mothers during the first year of life exhibited higher levels of distress and disturbance than the comparison groups. At two-and-a-half years the animals were again socially tested, and those repeatedly separated from their mothers were still subordinate to all others. Thus, the procedures employed in some laboratories, involving repeated separation of infants from their mothers for measurement, weighing, blood draws, etc., can have lasting consequences on behavior. The consequences are not necessarily *bad*, but it must be recognized that such seemingly innoccuous manipulations may influence later behavior.

Effects of proximal caging. While few data have been reported on the factors that may influence behavior of animals placed into newly formed groups (see Chapter 6, this volume), Ruppenthal and Sackett (in press) reported an interesting result of grouping macaques after the animals had been confined for a substantial period in individual cages. The animals were grouped so that they were with monkeys with whom they were familiar from being housed in adjacent cages, or with animals with whom they were unfamiliar. Aggression levels and trauma were much higher in groups formed from animals that had been housed near each other than those that were totally unfamiliar. This response is interesting in the light of other observations

(Chapters 3 and 6, this volume). Potential hazards of this type must be recognized in decisions on formation of groups.

Abusive mothers. Mitchell et al. (1967) and Sackett (1968) documented extremely high levels of aggression of offspring of abusive mothers. Thus, there is a potential for extragenetic transmission of aggressive behavior in colonies originally composed of females whose rearing histories resulted in development of hyperaggressivity. Fortunately, abusiveness tends to disappear across additional births if mothers receive sufficient experience with their infants (Ruppenthal et al., 1976).

CONCLUSIONS

Many of the results reported in this chapter and those subsequent to it were not anticipated. Clinical assessment needs basic information upon which to base prediction. The first level of information may be generalization from results of studies of other species, even nonprimates. The second, and preferred level, is data on the species in which one is interested or, at least, from congenerics. The third level, even more superior, is comparative data on the species in question across laboratories or housing conditions. The best evidence usually available is of the latter kind. Some of the research reported in this volume transcends these three levels in various ways. Some of the studies involve many comparisons within the laboratory using identical methods; others involve comparison of more than one species under similar circumstances.

We hope the future will bring about an even higher quality of comparison. Direct comparisons of captive groups under various environmental circumstances should be made using identical methods. Comparisons of many groups in captive breeding situations, in free-ranging natural circumstances, and across many zoo and laboratory conditions will provide the data essential for discovery of general principles of environmental influences on primate behavior. It is our hope that this volume will provide an impetus to continued and increased study of environmental influences on behavior and development of primates and other animals.

REFERENCES

Baumeister, A. and Forehand, R. Stereotyped acts. In *International Review of Research in Mental Retardation*, Vol. 6. New York: Academic Press, 1973, pp. 55–96.

Berkson, G. Abnormal stereotyped motor acts. In J. Zubin and H. Hunt (Eds.) *Comparative Psychopathology–Animals and Human.* New York: Grune & Stratton, 1967, pp. 76–94.

Berkson, G. Development of abnormal stereotyped behaviors. *Dev. Psychobiol.* **1**: 118–132 (1968).

Bernstein, I. Daily activity cycles and weather influences on a pigtail monkey group. *Folia Primatol.* **18**: 390–415 (1972).

Castell, R. and Wilson, C. Influence of spatial environment on development of mother-infant interaction in pigtail monkeys. *Behaviour* **39**: 202–211 (1971).

Chamove, A. The effects of varying infant peer experience on social behavior in the rhesus monkey. M.A. thesis, University of Wisconsin, Madison, 1966.

Davenport, R. and Menzel, E. Stereotyped behavior of the infant chimpanzee. *Arch. Gen. Psychiatry* **8**: 99–104 (1963).

Draper, W. and Bernstein, I. Stereotyped behavior and cage size. *Percept. Mot. Skills* **16**: 231–234 (1963).

Erwin, J. and Mitchell, G. Initial heterosexual behavior of adolescent rhesus monkeys (*Macaca mulatta*). *Arch. Sex. Behav.* **4**: 97–104 (1975).

Erwin, J., Mitchell, G., and Maple, T. Abnormal behavior in non-isolate-reared rhesus monkeys. *Psychol. Rep.* **33**: 515–523 (1973).

Gluck, J. and Sackett, G. Frustration and self-aggression in social isolate rhesus monkeys. *J. Abnorm. Psychol.* **83**: 331–334 (1974).

Griffin, G. The effects of multiple mothering on the infant-mother and the infant-infant affectional systems. Ph.D. dissertation, University of Wisconsin, Madison, 1966.

Harlow, H. F. The nature of love. *Am. Psychol.* **13**: 673–685 (1958).

Harlow, M. K. Nuclear family apparatus. *Behav. Res. Methods Instrum.* **3**: 301–304 (1971).

Jensen, G. and Bobbitt, R. Changing parturition time in monkeys (*Macaca nemestrina*) from night to day. *Lab. Anim. Care* **17**: 379–381 (1967).

Jensen, G., Bobbitt, R., and Gordon, B. Dominance testing of infant pigtailed monkeys reared in different laboratory environments. In *Proc. 3rd Int. Congr. Primatol.*, Vol. 3. Basel: S. Karger, 1971, pp. 92–99.

Kuo, Z. Y. *The Dynamics of Behavioral Development: An Epigenetic View.* New York: Random House, 1967.

Kuyk, K., Dazey, J., and Erwin, J. Play patterns of pigtail monkey infants: Effects of age and peer presence. *J. Biological Psychol.* **18**: 20–23 (1977).

Martenson, J., Oswald, M., Sackett D. and Erwin, J. Diurnal variation in common behaviors of pigtail monkeys (*Macaca nemestrina*). *Primates* **18**: 875–883 (1977).

Mason, W. and Berkson, G. Effects of maternal mobility on the development of rocking and other behaviors in rhesus monkeys: A study with artificial mothers. *Dev. Psychobiol.* **8**: 197–211 (1975).

Mitchell, G. Paternalistic behavior in primates. *Psychol. Bull.* **71**: 399–417 (1969).

Mitchell, G. Abnormal behavior in primates. In L. Rosenblum (Ed.) *Primate Behavior: Developments in Field and Laboratory Research*, Vol. 1. New York: Academic Press, 1970, pp. 195–249.

Mitchell, G., Arling, G., and Møller, G. Long-term effects of maternal punishment on the behavior of monkeys. *Psychonom. Sci.* **8**: 209–210 (1967).

Mitchell, G., Harlow, H., Griffin, G., and Møller, G. Repeated maternal separation in the monkey. *Psychonom. Sci.* **8**: 197–198 (1967).

Paulk, H., Dienske, H., and Ribbens, L. Abnormal behavior in relation to cage size in rhesus monkeys. *J. Abnorm. Psychol.* **86**: 87–92 (1977).

Riesen, A., Perkin, M., and Struble, R. Open-field behavior of socially deprived stumptail monkeys. Paper presented at the Inaugural Meeting of the American Society of Primatologists, Seattle, Wash., 1977.

Ruppenthal, G. and Sackett, G. Experimental and husbandry procedures: Their impact on development. In G. Ruppenthal (Ed.) *Nursery Care in Nonhuman Primates.* New York: Plenum Press, in press.

Ruppenthal, G., Arling, G., Harlow, H., Sackett, G., and Suomi, S. A 10-year perspective of motherless-mother monkey behavior. *J. Abnorm. Psychol.* **85**: 341–349 (1976).

Sackett, G. Abnormal behavior in laboratory-reared rhesus monkeys. In M. Fox (Ed.) *Abnormal Behavior of Animals.* Philadelphia: Saunders, 1968, pp. 293–331.

Sackett, G. and Ruppenthal, G. Development of monkeys after varied experiences during infancy. In *Ethology and Development.* Philadelphia: Lippincott, 1973, pp. 52–87.

Schein, M. W. and Hafez, E. S. E. The physical environment and behaviour. In E. S. E. Hafez (Ed.) *The Behaviour of Domestic Animals.* Baltimore: Williams & Wilkens, 1969, pp. 65–84.

Struble, R. and Riesen, A. Some effects of social isolation on somatic development in stumptail monkeys. Paper presented at the Inaugural Meeting of the American Society of Primatologists, Seattle, Wash. 1977.

Suomi, S. Adult male-infant interactions among monkeys living in nuclear families. *Child Dev.* **48**: 1255–1270 (1977).

Tinklepaugh, O. The self-mutilation of a male *Macacus rhesus* monkey. *J. Mammal.* **9**: 293–300 (1928).

Vernberg, F. and Vernberg, W. *The Animal and the Environment.* New York: Holt, Rinehart and Winston, 1970.

2
Primate Psychology in Historical Perspective*

Terry L. Maple

Department of Psychology, Georgia Institute of Technology, and Yerkes Regional Primate Research Center, Emory University Atlanta, Ga.

INTRODUCTION

Primate behaviorists like to portray their discipline as a new one. Indeed, the field has so rapidly advanced during the past two decades that it appears to have sprouted from nothing. Much of what we know today, however, is built upon the firm foundation of earlier traditions of inquiry.

The history of mankind's fascination with the other members of the primate order is a rich and diverse record of achievement. Unfortunately, a new generation of primate workers, in the laboratory and the field, know little of the history of primate behavior research. In the typical graduate curriculum these days, there is little concentration on the *classic* contributions of animal psychologists. To know the literature of the "old school" one must dig through the archives. Most of the important contributions of the early animal behaviorists are difficult to obtain, and there exist few adequate secondary sources. Primate behavior received attention from many early animal psychologists, and has been—with varying degrees of accuracy—descibed by many naturalists and explorers of previous centuries.

The present contribution is intended to be an overview of the achievements, methods, and focus of primate research as it has spanned the past several centuries. This effort must be brief, since the literature is voluminous. My main objective in outlining this

---

*The author gratefully acknowledges the assistance of Ryan Tweney (of *De Anima: Books in Psychology* and Bowling Green State University) who originally located many of the older volumes from which much of this paper has been drawn. William A. Mason has also been most helpful in pointing out references and areas of importance which were overlooked in the early stages of this manuscript. Harriett Powell added useful criticism of the style and G. Mitchell provided support and feedback which was most valuable. This paper is based on a symposium contribution to the Western Psychological Association, April 1975.

topic is to contrast the wisdom and the folly of the pioneer students of primate behavior. The material presented here is selective, but is fully representative, I think, of the spirit of the respective eras which are covered. I have been especially eager to counter the prevailing view among many field workers that early American psychologists were ignorant of the value of *naturalistic* research. As the reader will observe nothing could be further from the truth.

In examining the material which follows, the reader will notice that the early primate workers were often well aware of the problems inherent in the study of primates. For example, many researchers discuss the importance of understanding the natural behavioral propensities of their subjects, and the difficulty of breeding and keeping captive primates. This historical overview should be useful in assessing the progress which has been made in the study of primates in captivity.

THE EUROPEAN NATURALISTS

From the very beginning of the European expansion into the unknown territories of Africa, Asia, and South America, explorers and their patrons took an active interest in the indigenous fauna. Many specimens were acquired for study by European scientists and some exotic animals were brought back alive. The monkeys and apes were particularly sought and were the objects of much discussion. Where possible, monkeys were brought back as pets or as objects of display in zoos and menageries. European anatomists were especially interested in the dissection of these animals for comparison to aspects of human structure. While the behavioral observations of these naturalists were often exaggerated, they could not fail to improve upon the knowledge of the ancients as exemplified by the work of Pliny (A.D. 23–79):

> Tailed species (of apes) have even been known to play at draughts, are able to distinguish at a glance sham nuts made of wax and are depressed by the moon waning and worship the new moon with delight. (Book LXXX, P. 151).

An early example of the inferences drawn from the observation of free-ranging primates is to be found in Wakefield (1811):

> We are apt to attribute to them (the apes) peculiar sagacity, from their comi-

> cal tricks; though it seems that many other animals far excel them in their wild state, as to their natural habits, and the policy of their communities. The satyrs . . . in heathen mythology, were symbols of every inordinate passion: the model from which they were delineated, was evidently taken from the ape tribe, as may be seen by the strong similitude of the countenances of the one to the other. Nor does the dispositions of these creatures falsify their physiognomy: many species are said to be fierce, ill-natured, malicious, revengeful, thievish, mischievous, and immodest; exhibiting a picture if it may bear that term, of man in the most debased condition, a slave to vice and his own unrestrained inclinations. (Pp. 228–229)

Regarding baboons, we learn:

> Their propensities and figure are altogether most disgusting. (P. 231)

This description, a common nineteenth century reaction to nonhuman primates, reminds me of a remark by the humorist Will Cuppy. In 1931 he wrote, "The baboon is entirely uncalled for. . . ." He added, "never call anyone a baboon unless you are sure of your facts." While many writer's attitudes toward primates were uncomplimentary, to say the least, others were downright silly, such as the popular view that apes abducted women.

Much of the early information about monkeys and apes was based on the haphazard anecdotes of hunters and explorers. Often these writers would obtain their information from the natives who inhabited the African, Asian, and South American regions. These tales may have been embellished by the natives to impress the Europeans, and further altered to *sell books*!

On the other hand, some of the books of the day were remarkably accurate. For example, Henry Forbes' 1896 *Handbook to the Primates* quotes a Dr. Holub:

> I was turning to leave the ravine when some stones came pattering down the rocks in my direction. I soon became aware that the stones were being designedly *aimed* at me; and, looking up, I saw a herd of baboons. (Vol. I., P. 256)

This description of the use of rocks as weapons by chacma baboons predates by seventy-nine years Hamilton's recent description of the same phenomenon (Hamilton et al., 1975). In the same book (Forbes, 1896) an accurate description of hand-slapping threat displays of baboons is provided. Forbes also knew of baboon predation, a subject of great interest to modern primate workers (see Strum, 1975).

It is interesting to note that some of what we know today about the expressive behavior of baboons was known in 1872. Darwin wrote:

> Baboons often show their passion, and threaten their enemies in a very odd manner, namely by opening their mouths widely, as in the act of yawning. Mr. Bartlett has often seen two Baboons, when just placed in the same compartment, sitting opposite to each other, and then alternately opening their mouths; and this action seems frequently to end in a real yawn. Mr. Bartlett believes that both animals wish to show to each other that they are provided with a formidable set of teeth, as is undoubtedly the case. As I could hardly credit the reality of this yawning gesture, Mr. Bartlett insulted an old baboon and put him into a violent passion; and he almost immediately thus acted. (*Expressions of the Emotions*, Pp. 136–137)

Darwin drew many parallels between nonhuman primates and *Homo sapiens*. He cited many examples of sexual dimorphism in monkeys to account for his theory of sexual selection, and Darwin's descriptions of facial expression and posture in monkeys and apes were remarkably accurate. An interesting quote scribbled into one of Darwin's notebooks (Gruber and Barrett, 1974) gives some insight into his views: "He who understands baboon would do more toward metaphysics than Locke."

G. J. Romanes, who has been called the chief anecdotalist of his day, also studied primates. A young protege of Darwin, he successfully taught a captive chimp to associate numbers with particular amounts of straw. He made the astute observation that the difficulty of learning more than five associations was probably due to the chimp's impatience rather than to any lack of intelligence. About this chimp he wrote in 1889:

> Her intelligence was conspicuously displayed by the remarkable degree in which she was able to understand the meaning of words—a degree fully equal to that presented by an infant a few months before emerging from infancy and, therefore, higher than that which is presented by any brute, so far at least, as I have met with. . . . (Pp. 316–317).

My personal view is that Romanes may have been a better observer and a better scientist than that for which he has been credited. It should be noted that Darwin, too, was a compiler of anecdotes, having been especially limited for most of his life by the lingering effects of a mysterious illness.

The early naturalists were mainly motivated by the desire to collect and classify animals. Thus, many specimens of primates were shot, dissected, catalogued, and put on display. Even so, many collectors found it difficult to shoot primates, reflecting on their resemblance to mankind. Forbes (1896) wrote:

> There is a natural repugnance to collecting specimens of monkeys on the part of sportsmen. To shoot one feels like killing a sort of relation, and even our best collectors, who thoroughly understand the necessity of obtaining specimens in the interest of science speak with a feeling of pain of the humanlike distress which a wounded monkey exhibits. . . . (Introduction)

Similarly, George Gulliver, writing in the journal *Nature* (1873), quoted the following anecdote from Forbes' *Oriental Memoirs* of 1813:

> One of a shooting party, under a banian tree, killed a female monkey and carried it to his tent, which was soon surrounded by forty or fifty of the tribe, who made a threat noise and seemed disposed to attack their aggressor. They retreated when he presented his fowling piece, the dreadful effect of which they had witnessed and appeared perfectly to understand. The head of the troop, stood his ground, chattering furiously; the sportsman, who perhaps felt some little degree of compunction for having killed one of the family, did not like to fire at the creature. . . . At length he came to the door of the tent, and finding threats of no avail, began a lamentable moaning, and by the most expressive gesture seemed to beg for the dead body. It was given him; he took it sorrowfully in his arms, and bore it away to his expecting companions: they who were witness of this extraordinary scene, resolved never again to fire at one of the monkey race. (P. 103)

While much of this early interest in primates was anatomical, the work of Darwin and other evolutionary theorists sparked an interest in the habits and "mental-life" of these animals. Many of the observations which were made by such naturalists predated later experimental work and are unfamiliar to many modern workers. For example, Alfred Russell Wallace, the co-founder of the theory of natural selection, was the first scientist to describe the construction of an inanimate surrogate mother. In *The Malay Archipelago* (1869) he described the construction of an "artificial mamma" for an orphaned orangutan baby out of a buffalo skin made up into a bundle, in the following manner: "At first this seemed to suit it admirably, as it could sprawl its legs about and always find some hair, which it grasped with the greatest tenacity."

In the same volume Wallace also discussed the comparative development of a young orangutan (Mias) and a young hare-lipped monkey (*Macaca cynomolgus*):

> It was curious to observe the different actions of these two animals, which could not have differed much in age. The Mias, like a very young baby, lying on its back quite helpless, rolling lazily from side to side, stretching out all four hands into the air, wishing to grasp something, but hardly able to guide its fingers to any definite object, and when dissatisfied, opening wide its almost toothless mouth, and expressing its wants by a most infantive scream. The little monkey, on the other hand, in almost constant motion; running and jumping about wherever it pleased, examining objects around it, seizing hold of the smallest objects with the greatest precision, balancing itself on the edge of a box or running up a post, and helping itself to anything eatable that came in its way. There could hardly be a greater contrast, and the baby Mias looked more babylike by the comparison. (Pp. 63-64)

THE TWENTIETH-CENTURY EXPERIMENTALISTS

During the early decades of the twentieth century, naturalists and adventurers continued to explore the wilds of Africa, Asia, and the Americas. However, the revolution in experimental techniques which affected all of psychology was especially influential in the area of comparative psychology. Warden (1927) provided an excellent discussion of the experimental developments in this subject area, but the experimental study of nonhuman primates was not included. During the first thirty years of the twentieth century, many important papers were published by individuals interested in primate psychology. Among these pioneers were John B. Watson (the founder of doctrinaire behaviorism), Wolfgang Köhler (an advocate of *gestalt* psychology), Robert Yerkes, and G. V. Hamilton. Other notable publications by prominent psychologists followed, leading up to the modern era.

Stimulus and Response

The first few papers on primate psychology published in this century concerned the sensory capacities and learning abilities of monkeys.

This research corresponded to the general scope of comparative psychology at that time. Darwin had opened the door for comparison of animals to humankind, and scientists were busily attempting to demonstrate that differences in structure and behavior were matters of degree. The monkeys and apes, because of their structural similarities to people were obvious links to our primate heritage.

In 1902, A. J. Kinnaman wrote about the *mental life* of captive rhesus monkeys in the *American Journal of Psychology*. Kinnaman's extensive paper (in two parts) was especially wide ranging, including remarks on methodology. Two direct quotes will make this point:

> One of the great difficulties is that of selecting tests and apparatus properly related to the character of the animal studied. . . . The tests selected must not be too far removed from the field of his natural activity, for in such cases the animal is often completely baffled, and only the most limited results are obtained. The animal may appear to be stupid because the test is unsuitable, and thus be easily misjudged. (P. 103)

> There is reason for believing that if all *methods*[1] are employed upon the same animal, as far as possible, we shall arrive at a more complete understanding of him. Not opposition to one method or undue emphasis upon another, but a reasonable use of each within the just limits of its applicability, should be the practice of the student of comparative psychology. (P. 103)

These pioneering views can be compared with the recent conclusions of Breland and Breland (1966):

> . . . you cannot understand the behavior of the animal in the laboratory unless you understand his behavior in the wild. (P. 20)

> Species come equipped to recognize or react to only certain inputs or portions of the environment. There can be no input without a certain impingement from the environment, and there can be no intake without an "accepting" process in the organism. This accepting process is definitely concerned with nervous structure and function, both with end organ (eye, ear, taste buds, and other receptive cells) and with central nervous system processes. . . . (P. 60)

Kinnaman, like Yerkes (as will be clear) understood the need to supplement laboratory observation with data from the natural habitat. He also was aware of the effects of captivity on behavior:

> . . . from a purely scientific standpoint also, this (naturalistic) method can be

---

[1] Italics are the author's (T. L. Maple).

made to bear first-class fruit. It has the advantage of seeing the real animal in his natural, unhampered reactions. (P. 99)

The caged animal often ceases to be himself. He varies from insipid tameness and moroseness to wild excitement, and is in several ways very different from what he is in his larger freedom. (P. 101)

Kinnaman was a first-rate experimenter whose views could be seen as midway between those who relegated primate behavior to the realm of "tropisms, instincts and associations" and those who ascribed to them "human" powers. After his research on the mental powers of rhesus monkeys, he wrote:

The sort of errors made, the elimination of many unnecessary movements and the character of many other reactions not classified here incline the writer to believe that the monkeys' mental processes are after all not so simple as analysts have often asserted them to be. Whether these animals have "free ideas" and general notions beyond the mere "recept," and are capable of real analogical reasoning cannot be positively determined. If they do, the processes certainly do rise to the level of full reflective consciousness. Yet there is no way of knowing, because there is no certain way of having the consciousness that the animal has. But that these monkeys have acted objectively just as human beings act when they have these mental activities is most certain. I am inclined to believe that the human and animal consciousnesses are not really different in kind but only in degree, the difference in degree, however, is very great. (P. 212)

With the 1901 publication "The mental life of the monkeys," Edward L. Thorndike, who contributed so much to the literature of comparative animal learning, is credited with being the first investigator to bring a monkey into the laboratory for experimental psychological research (Kinnaman, 1902). His work concerned the intelligence of three *Cebus* monkeys, with which he experimented for several months. At this time, the *Cebus* was thought to be the most intelligent of all monkeys, a not-so-surprising belief since *Cebus* were well known as the clever and adaptable partners of organ-grinders throughout Europe and America. What Thorndike found was that none of the monkeys solved puzzle-box problems in a fashion which could be interpreted as "reason."

Thorndike concluded that monkeys learned, as do cats and dogs, by trial and error. However, he also believed that monkeys learned in a more efficient manner. Monkeys were thought to form more habits, according to Thorndike, and these habits were said to be formed more quickly and were longer lasting than in cats and dogs.

A male capuchin (*Cebus* sp.) belonging to De Haan (1931) when tested by most of the methods utilized by Köhler, convinced De Haan that it was no less "intelligent" than the chimp, and more so than the gorilla, orang, or gibbon. Zuckerman (1933), in discussing this finding, wrote that ". . . there is every reason to suppose that some catarrhine monkeys, if tested in the same experiments, would yield results as good as those of De Haan's capuchin."

Klüver (1933) has also been closely associated with the search for intelligent behavior in nonhuman primates. His work on the learning and perceptual abilities of monkeys is remarkably detailed, and he was especially impressed with the performance of *Cebus* monkeys. About the capacity of monkeys to reason, he wrote:

> It is true that the monkey does not develop such a thing as the Kantian philosophy; at any rate he does not write a "Critique of Pure Reason," but it may be asserted that he is able to think categorically. The fact that the material he handles when thinking is different from the material human thinking is concerned with, does not prove that the monkey cannot think "categorically" at all. (P. 360)

On the other hand, Harlow et al. (1932) and Maslow and Harlow (1932) found capuchins to be decidedly inferior to baboons, macaques, gibbons, and even lemurs in tests of delayed reaction. Zuckerman (1933) explained the confusion surrounding the differential abilities of monkeys and apes by referring to the earlier conclusion of St. George Mivart (1874):

> Yet the psychical powers of different Apes are very similar. Not only the lowest baboons of Africa (as e.g., the famed "Happy Gerry" of Exeter charge) can be taught various and complex tricks, but the less manlike American monkeys —the common sapajous (cebus)—are habitually selected by peripatetic Italians for the exhibition of the most clever and prolonged performances.
>
> As to the two species of sapajous, the brains of which are so different the one from the other, Professor Rolleston asks: "Will anybody pretend that say difference can be detected in the psychical phenomena, the mental manifestations of these creatures, at all in correspondence or concomitant variation with their differences of cerebral conformation?"
>
> The difference between the brain of the Orang and that of Man . . . is a difference of absolute mass. It is a mere difference of degree and not of kind.

> Yet the difference between the mind of Man and the psychical faculties of the Orang is a difference of kind and not one of mere degree.
>
> Thus on the one hand we see that we may have great differences in brain development unaccompanied by any corresponding psychical diversities, and on the other we may have vast psychical differences which it seems we must refer to other than cerebral causes. (Pp. 148–149)

One of the central problems which occupied the time of primate workers was this difficulty in comparing different species. Klüver was especially aware of these problems:

> Suppose we give the same "intelligence" test to different monkeys and find that monkeys in a certain species cannot succeed in it. An analysis might show then the monkeys of this species cannot even see the objects presented. It is not . . . sufficient to present certain stimulus situations, we must find out what the situation is in terms of the animal. (P. 357)

A factor involved in the differential abilities of various species was the difference in attention. Furthermore, individual differences could account for some of the conflicting results of experimenters accustomed to using small numbers of monkeys in their research. C. Lloyd Morgan (1891) who did little research with monkeys, understood this quite well:

> Animals differ widely in their power of attention, as every one knows who has endeavoured to educate his pets. Darwin tells us that those who buy monkeys from the Zoological Gardens, to teach them to perform, will give a higher price if they are allowed a short time in which to select those in which the power of attention is more developed. (P. 342)
>
> Everyone must have seen animals defining their constructs by examination. A monkey will spend hours in the examination of an old bottle or a bit of looking glass. At the Zoological Gardens connected with the National Museum at Washington, a monkey was observed with a female opossum on his knee. He had discovered the slit-like opening of the marsupial pouch, and took out one and then another of the young, looked them over carefully, and replaced them without injury. (P. 340)

Harry Harlow is best known for his work in the fifties and sixties, but he also contributed heavily to the earlier literature on primate behavior. He began his Wisconsin studies in 1930 and counted among his first graduate students A. H. Maslow. Harlow and Maslow (for lack of available space) began their primate work at the Vilas

Park Zoo in Madison. Hahn (1971) quotes Harlow's reminiscences about the early days:

> We gave him (Jiggs the orang-utan) two oak blocks, one with a square hole and one with a round hole, and a square plunger and a round plunger. He learned to put the round plunger in the round hole and the round plunger in the square hole, and he learned to put the square plunger in the square hole, but he never learned to put the square plunger in the round hole. He worked incessantly on this unsolvable problem for six months and then died of perforated ulcers, but at least he died demonstrating a level of intellectual curiosity greater than that of many University of Wisconsin students. (P. 74)

After moving to an abandoned building on the Wisconsin campus, better controlled studies of monkey behavior became possible. Three major thrusts of research resulted: the development of measures of nonhuman primate learning, the effects of cortical lesions on these learned behaviors, and the analysis of primate motives. Harlow gained in stature with his work on learning set, summarized in major publications in 1949 and 1951. In 1959, the Wisconsin Primate Laboratory was institutionalized during which time Harlow's famous studies of development and socialization were taking place. A future historian of primate research will doubtless note that Harlow influenced more students than any primate behaviorist since Yerkes.

Wolfgang Köhler and the Demonstration of Insight

Köhler was not only an important figure in the development of a "primate psychology," but a leader in the movement known as *gestalt* psychology. His most important work, as far as primate behavior is concerned, was carried out on the Mediterranean island of Tenerife during the years 1913–1917. Köhler was interested in the demonstration of higher intelligence in the chimpanzee. He, too, was well aware of the fact that an important barrier to the study of anthropoid intelligence was the experimenter (Köhler, 1925):

> . . . the success of the intelligence tests in general will be more likely endangered by the person making the experiment than by the animal . . . every intelligence test is a test, not only of the creature examined, but also of the experimenter himself. (P. 265)

Köhler believed that a primary difference between man and chimpanzee was the latter's limited capacity to perceive the past and the future. The key concept in Köhler's investigation was *insight.* He specifically aimed to demonstrate that apes behaved with insight under conditions which required it, and that this behavior was comparable to that exhibited by human beings. Insight was thought to indicate *intelligence*:

> ... we tend to speak of "intelligence" when, circumstances having blocked the obvious course, the human being or animal takes a round about path, so meeting the situation. (Pp. 3-4)

Köhler observed that the importance of demonstrating insight in chimpanzees is that it would revise current notions of animal intelligence, since prior studies failed to utilize the "higher" animals such as the anthropoid apes. His method of operation was to block the obvious path to an objective, leaving only a *roundabout* solution. He used nine chimpanzees in his studies, ranging from four to seven years of age, comparing these directly to humans, hens, and one dog.

During the course of his studies, Köhler made important observations regarding individual differences in chimp temperament as well as learning ability. Köhler's results were most significant, in that the extraordinary problem-solving capacity of the chimpanzee was amply demonstrated. Subsequent work supported Köhler's contention that the chimpanzee was a remarkably intelligent animal.[2] McDougall and McDougall (1931) attempted to support the ideas of Köhler and refute the earlier views of both Thorndike (1898) and Lloyd Morgan (1929). He argued further that

> ... Köhler, in breaking through the bounds set by the old association theory (still dominant in animal psychology), does not go far enough; that we must recognize that animal behavior is guided not only by insight (the grasping of essential relations, such as those of time, space and causation) but also by foresight, foresights of relations, not yet realized or presented, but to be realized by the agency of the movements which the animal is about to make; further that such insight and foresight are not confined to the higher apes, but are unmistakably manifested by common monkeys, by raccoons, by rats, and by wasps. (P. 239)

[2] Köhler was also aware of the need to supplement laboratory work with naturalistic studies. He wrote: "The results of special experimental investigation only take on the real colour of life when the habits and the character of the creatures under observation have become adequately known in the *natural expression.*" (Emphasis is the author's—T. Maple.)

Development and Social Behavior

From the very beginnings of primate behavior research, students recognized the importance of social development and social behavior. Some information in this regard had been passed on by the naturalists such as Wallace and Cuvier. An obscure but reasonably astute example is the following passage from Ram Bramha Sanyal (1892) concerning the development of an infant macaque in the wild, as cited in Foley (1934):

> The young monkey after birth attaches itself to its mother, and will not leave her for nearly a month, the mother nursing the young all the time with the utmost solicitude; after this time it will make little excursions on its own account, but is careful not to stray far, and at the slightest sound or movement it seeks refuge with her. The mother is unremitting in her vigilence over her offspring and in its personal wants and appearance. Compared with an orang-utan of the same age, a monkey is more helpful and intelligent, and in fact all its instincts are strongly developed at a comparatively early age. In about a month the young one begins to pick up grain and other food, and the struggle for life soon begins, and the mother and the young one commence to fight over the food, although their natural instincts bind them to each other at other times. (P. 45)

An interesting paper published in 1879 concerned the reaction of a young chimpanzee to the loss of a cagemate. The writer, A. E. Brown, in describing the behavior of this animal, illustrates a phenomenon which primate workers now refer to as the *Bowlby syndrome*:

> After the death of the female . . . the remaining one made many attempts to rouse her, and when he found this to be impossible his rage and grief were painful to witness . . . he made repeated efforts to arouse her, lifting up her head and hands, pushing her violently and rolling her over. After her body was removed from the cage–a proceeding which he violently opposed–he became more quiet, and remained so as long as his keeper was with him . . . the day following, he sat still most of the time and moaned continuously. . . . (P. 174)

This early description resembles the pattern described by Bowlby (1968) and others (Seay and Harlow, 1965; Scollay, 1970) in both humans and monkeys when separated from a *loved one*; the first phase of protest, and the second phase of despair. Goodall (1968) recently observed in the wild a similar pattern in a young chimp which lost its mother. The ultimate death of this animal is thought to have been triggered by grief at the loss of the mother.

Lashley and Watson (1913) also contributed valuable information on the development of behavior in the infant monkey. Their paper was a diary of observations which compared the development of the young macaque to that of the young human child. Lashley and Watson discussed learning, play, and mother-infant interactions. They were surprised to find a short infancy in the monkey, a finding which agreed with the earlier observation of Wallace.

Watson (1908) contributed other data on social behavior as well as his primary interests in learning. Regarding grooming (flea-catching) he wrote:

> Flea-catching, regardless of what the sociologist may have to say, is the most fundamental and basal form of social intercourse between rhesus monkeys.[3]

Foley (1934, 1935) made some outstanding contributions to the literature, much of which is directly applicable to current studies of primate development (cf. Mitchell, 1970), particularly his work on *isolated* monkeys. In these latter publications he describes abnormal swaying movements, thumbsucking, self-grasping, and lack of sexual behavior in his subject. Because the infant survived the isolation experience, Foley was convinced that this species could be most useful in further studies:

> . . . we can look forward to the future utilization of the infant *Macaca mulatta* monkey in an attack upon many of the intriguing programs of psychology. . . . (P. 50)

Early research on sexual behavior was influenced by the writings of Freud, Adler, and other psychoanalytic writers. For example, Hamilton, Maslow, and Zuckerman utilized psychoanalytic constructs in their discussions of sexual aberrations. A second influence in this was the interest in the successful husbandry of apes. It was thought that a proper understanding of the natural pattern of sexual behavior would lead to improvements in reproductive success. Consequently, many descriptions of sexual relations are to be found in the early lit-

---

[3]Tinklepaugh (1931), Yerkes (1933), and Zuckerman (1933) also wrote at length on grooming behavior.

erature. Yerkes, as in other areas of primate behavior, was a principal authority on the reproductive behavior of nonhuman primates. Extensive details regarding great ape husbandry are found in his books *Almost Human* (1925), *The Great Apes* (1929), and *Chimpanzees: A Laboratory Colony* (1943).

G. V. Hamilton's (1914) work on the sexual behavior of monkeys and apes stemmed from a desire to test Freudian hypotheses on animal subjects, particularly as it concerned homosexuality and other "abnormal behaviors." He formulated two basic questions:

> 1) Are there types of infrahuman primate behavior which cannot be regarded as expressions of a tendency to seek sexual satisfaction but which have the essential objective characteristics of sexual activity?
> 2) Do such sexual reaction types as homosexual intercourse, efforts to copulate with nonprimate animals, and masturbation normally occur among any of the primates, and, if so, what is their biological significance?

Hamilton concluded that sexual behavior was motivated by "hunger for sexual satisfaction, hunger for escape from danger, and, possibly, hunger for access to an enemy." He added that:

> . . . homosexual behavior is normally an expression of tendencies which come to expression even when opportunities for heterosexual intercourse are present. . . . Masturbation does not seem to occur under normal conditions. The macaque of both sexes is apt to display sexual excitement in the presence of friendly or harmless non-primates. It is possible that the homosexual behavior of young males is of the same biological significance as their mock combats. It is clearly of value as a defensive measure in both sexes. Homosexual alliances between mature and immature males may possess a defensive value for immature males, since it insures the assistance of an adult defender in the event of an attack. (P. 317)

Maslow's series of studies published in the mid-thirties, established that dominance was a factor in the sexual life of monkeys. He separated sexual behavior into two distinct categories, that motivated by the *sexual drive* and that motivated mainly by the *dominance drive.* Maslow's *two-drive theory* was called into play to account for various idiosyncrasies observed in nonhuman primates such as prostitution behavior, homosexual behavior, lack of sexual interest, sadism-masochism, jealousy, and ventroventral postures.

Robert M. Yerkes

The history of primate research in this century is closely linked to the accomplishments of Robert Mearns Yerkes.[4] His direct influence spanned the first fifty years and continues to be expressed in the work of those he trained and with whom he collaborated. Known primarily for his work with primates, Yerkes was a first-rate comparative psychologist who studied many diverse species under a variety of conditions. His early contributions included apparatus for the study of visual processes and learning. Yerkes was not only an inventive and hard-nosed researcher, but a planner as well. His dream of a laboratory in *psychobiology* was formulated during his graduate training at Harvard.

> A plan whose realization after nearly thirty years has now been nearly achieved in Yale University, came to me as a stirring vision of usefulness during my graduate days in Harvard. It was the establishment and development of an institute of comparative psychobiology in which the resources of the various natural sciences should be used effectively for the solution of varied problems of life. (Yerkes, 1932, P. 390)

In contributing some early studies of monkeys (Yerkes, 1915, 1916), he was preoccupied with the desire to investigate the intellectual capacities of great apes. In this desire he was not alone, and his early thinking was influenced by many of his contemporaries such as Köhler, Hamilton, and Kohts.

G. V. Hamilton, who had been a student of Yerkes, had established a laboratory in Montecito, California, where he pursued his interests in nervous and mental disorders.

> Hamilton . . . had the novel idea of using monkeys to illuminate problems of human behavior. His success was extraordinary. An investigator with lively speculative inclinations, his study of primate behavior was motivated by an intense interest in the problems of behavior and experience. His critical observations, conclusions, advice, and encouragement greatly helped to perfect my plan, confirmed my faith in the importance of the ideas behind it, and reinforced my efforts to arouse interest in its realization (Yerkes, 1943, P. 293).

Yerkes spent a summer working with Hamilton and studying the learning ability of his lone orangutan *Julius*. During this time he con-

---

[4] Much of what appears here is discussed in greater detail in Hahn (1971).

tinued to correspond with Köhler and made plans to visit the Cuban breeding facility of Rosalia Abreu. This visit was eventually arranged and took place in 1924. Yerkes was most impressed with Madame Abreu's success in anthropoid care and husbandry and described her remarkable facility in his book *Almost Human* (1925). In this book Yerkes made a remarkable statement which bears consideration in light of the recent work of Gardner and Gardner (1969), Fouts (1973) and others.

> I am inclined to conclude from the various evidences that the great apes have plenty to talk about, but no gift for the use of sounds to represent individual, as contrasted with racial, feelings and ideas. *Perhaps they can be taught to use their fingers, somewhat as does the deaf and dumb person, and thus helped to acquire a simple nonvocal, "sign language."*[5]

Although the First World War interrupted Yerkes' progress in psychobiology (but not his contributions to psychology), he eventually elaborated on the first published version of his dream "Provisions for the Study of Monkeys and Apes," which appeared in 1916.

By observing the success of Hamilton, Abreu, Köhler, and the workers of the Pasteur Institute's field station in Africa, Yerkes gained confidence in the efficacy of his dream. He was also encouraged by the experiments of Ladygin Kohts (1928, 1930) in the Soviet Union. She had conducted detailed studies of sensory and perceptual processes of a young chimp and a human child. Her work attracted the attention of Soviet biologists, leading to the establishment of the Sukhumi primate center in 1927 (Yerkes, 1943).

In 1925, Yerkes succeeded in obtaining the needed funds from Yale University and the Rockefeller Foundation. The promise of this financial commitment lured Yerkes from governmental work in Washington which had occupied his time since the end of the war. His plan was elaborated as follows:

> . . . (1) a general headquarters laboratory to be located in New Haven, in which a group of immature chimpanzees should be available as experimental subjects; (2) a breeding station, suitably located, which should maintain a colony of these apes, bred and reared under as nearly ideal conditions as practicable, and

---

[5] Emphasis is the author's (T. Maple). It is interesting to note that in the 1970s, having abandoned introspection through the teachings of Watsonian behaviorism, we are finally on the verge of asking an ape to tell us "how it feels to be an ape."

kept primarily for general observational use and as material for problem solution; and finally (3) a field study division, under which, as opportunity appeared, visits or expeditions to Africa could be arranged in order that the animals might be studied in their natural environment." (Yerkes, 1943, P. 296)

The field station of the *Primate Laboratory of the Yale Institute of Psychobiology* was established in 1930, coincident with the death of Madame Abreu. As a result of this event, and Yerkes' longtime friendship, Madame Abreu's children presented him with thirteen chimpanzees. Sixteen others followed in the next few years, brought from Africa by Henry Nissen as a gift from the Pasteur Institute. Yerkes' early anthropoid studies were wide ranging in scope. He maintained an active interest in husbandry, expression, locomotion, sexual behavior, attention, insight, memory, and adaptability. In his learning studies, Yerkes equated *Ideation* with insight, but on the whole he was somewhat less successful than Köhler. He did succeed, however, in gathering data on the gorilla and the orangutan to compare with that on chimps.[6]

Some impressive data on Yerkes' influence have been compiled. In 1968 a review of the literature on chimpanzees revealed that only one-fourth of the total number of references to that date occurred prior to the appearance of Yerkes' name in the literature (1916). His influence was felt in virtually every facet of research. Since his earlier work, Yerkes' name appears on fifty-six papers concerning the chimpanzee, for a total of 2,221 pages of material. He directly influenced 119 productive scientists including Nissen, Crawford, Zuckerman, Riesen, Spence, and others. Yerkes' students and co-workers produced 486 references in the chimp literature from 1916 to 1970. One of his greatest works, coauthored by his wife Ada Watterson Yerkes, *The Great Apes* (1929) was hailed by the *New York Herald Tribune* as "the greatest book on the subject ever written . . . an achievement that approaches the incredible." It may still be so labeled today. As Carmichael (1968) has written:

> Yerkes did much for comparative psychology in general but his greatest contribution was in developing the modern psychobiology of the infrahuman primates (P. 62).

---

[6] Will Cuppy (1931) has also commented on comparing the behavior of great apes: "If the scientist places a banana in a box the *chimpanzee* will go and get it and eat it . . . when a banana is placed in a trick box within easy reach, the *gorilla* will bite the professor's cousin. . . ."

THE FIELD WORKERS

Yerkes' original plan, as we have seen, included the provision for field work. The first of these studies was published by Henry Nissen in 1931, and concerned (appropriately enough) the social and ecological relations of the chimpanzee. This monumental study was quickly followed by Harold Bingham's (1932) study of the mountain gorilla, a study which, unfortunately, yielded little data on the social relations of these animals. As important as these first two efforts proved to be, the most influential of Yerkes' disciples was C. Ray Carpenter, who carried his psychological training into the field in 1933. His published paper on the howling monkeys of Panama (1934) is a classic contribution to the literature and was followed by field studies on spider monkeys, orangs, gibbons, and rhesus monkeys as well as by observational research on captive gorillas (cf. Carpenter, 1964). Carpenter's psychological background is especially apparent in his own words:

> I realized during the early field studies that the processes of learning, conditioning, and adaptive behavior operate in naturalistic behavioral systems. It could be inferred, also, that genetic factors determine the possible modifications of behavior of primates living in "free-ranging" groups and populations. Attempts have been made, therefore, to use the concepts of learning, conditioning, and development as elementary and partial explanations for the behavior that is characteristic of primates living in natural habitats. It was clear that to separate genetically or phylogenetically determined behavior from learned behavior was impossible. The concept of naturalistic behavior has been used to avoid this dichotomy. (P. 478)

Yerkes (1943) described this work as ". . . the first reasonably reliable working analysis of the constitution of social groups in the infra-human primates, and of the relations between sexes and between mature and immature individuals for monkey or ape."

He added this prophetic note:

> Carpenter's contributions may be counted on to command the attention and stir the enthusiasm of other investigators. Probably it will induce them to go and to try to do likewise.

This view is echoed by Southwick (1963) in retrospect:

> It would be difficult to overemphasize the importance of Carpenter's endeavors of the 1930's in terms of their impact on subsequent field studies. These

works established many of the motivations, goals, methodologies, and basic concepts of the subsequent work in their field. His monographs on the howlers of Panama and the gibbons of Thailand, and his review article in *Biological Symposia* of 1942, remain as classics in the field. (P. 2)

In 1946, J. P. Scott, C. R. Carpenter, T. C. Schnierla, F. Beach, and others organized a *Committee for the Study of Animal Societies Under Natural Conditions.* This group organized for many years interdisciplinary meetings and symposia with the *American Society of Zoologists* and the *Ecological Society of America*, later becoming the *Animal Behavior Society.*

THE MODERN ERA

Yerkes retired as director of the Yale labs in 1942 and was succeeded by the famous neurophysiologist, Karl Lashley. Lashley was especially aware of developments in Europe and, as director, was instrumental in bringing ideas from abroad into focus in America. Unfortunately for the progress of primate behavior the Second World War interrupted much of the promising work of the thirties. Although some work continued, most notably at Orange Park, progress was greatly slowed. The decade of the 1950s, however, was a time of rapid growth. Of particular importance was the work which was initiated by the Japanese scientists Iminishi, Miyadi, Itani, Kawamura, Kawai, Mizuhara, and others. This work carried out on free-ranging groups of *Macaca fuscata* is the most complete and lengthy record of primate field work ever conducted. A most notable finding has been the observation of the transmission of a novel idea from one member to the entire group, what has become known as *protoculture* (cf. Kawai, 1965).

The sixties were even more lively with a wealth of field and laboratory studies, and the establishment of the seven United States primate centers which continue to improve upon the models of the past. In addition to this promising data-based work, a growing awareness of the need for primate conservation was developed (cf. Bermant and Lindburg, 1975). There are now four major journals devoted to primate work, and many other more general periodicals in which primate studies appear. Indeed, one of the great problems of the modern era is keeping in touch with the growth of the literature.

CONCLUDING REMARKS

Despite these many valuable written accounts, primate studies did not really come of age until the twentieth century. Prior to this time, many factors prevented its development. Primates have always been difficult to keep and breed in captivity. Without the necessary expertise to maintain monkeys and apes, laboratory work could not progress, hence the concentration by so many early workers on the behavioral aspects of reproduction and development. Another important impediment to the field was the prevailing public attitude:

> The monkey has for centuries been a "cursed" animal. Its anatomical similarity to man . . . seemed a reproach to the authors and advocates of the biblical version of the divine creation of man. . . . It can be well supposed that had it not been for the zeal of the Holy Inquisition which destroyed essays on man's kinship with the monkey, and oftentimes the authors along with the essays, much more information on this question would have come down to us from the Middle Ages. Getting a scientific knowledge of monkeys had a direct bearing on the fundamental problems of man's origin, and was thus in the focus of the bitter fight between idealism and materialism. The Inquisition sent Giordano Bruno and Lucile Vanini to the stake; even in the XVIII century, ecclesiastics exerted pressure on the French naturalist Georges Buffon who classified monkeys and human beings as "quadrumanes" and "bimanes." (Lapin, cited in Hahn, 1971, P. 3)

There is much to be gained from the earliest observations of naturalists, despite the silly errors and misconceptions which are so often represented in discussions of the history of behavioral science. In the twentieth century, as we have seen, the pioneers of comparative psychology, particularly those working with primates, were more astute than many modern workers might admit. A rereading of classic papers can give us insight into problems which we struggle with today. We may avoid unnecessary repetition by these efforts and we cannot help but to gain a healthy appreciation for the wisdom, patience, and problems which beset those who paved the way for our contemporary efforts. It should be added here, though I loathe to submit to pessimism, that with money scarce, historical research is an economical way to support a hard-dying habit for primate work. Bad times or good, we are nonetheless likely to discover ever more valuable principles of behavior from the efforts of our present and future work on primate history. In so doing, it is altogether fitting

Fig. 2-1. Alfred Russell Wallace made early observations on the development of young monkeys and orangutans.

Fig. 2-2. Wolfgang Köhler's early studies of anthropoid ape intelligence also provided useful information concerning their reactions to a captive life.

Fig. 2-3. Robert M. Yerkes, with Chim and Panzee, gave considerable emphasis to the needs of captive primates.

Fig. 2-4. The gorilla Congo was an early subject of Yerkes' research. Here she contemplates her image in a mirror.

to consider the words of Yerkes (1932), the true founder of the field of primate behavior:

> Looking backward over thirty years of diligent labor and abundant intellectual, social, and material rewards, I am impelled to view all as preparation for the future. It is as if I were now on the threshold of a great undertaking which from the first was dimly envisaged and later planned for with increasing definiteness and assurance. Whether in this characterization of my past and prophecy for my future I am substantially correct, time will reveal. As ever, I am

Fig. 2-5. N. Ladygin Kohts was one of first workers to successfully keep a chimpanzee in captivity.

Fig. 2-6. Harry F. Harlow, with rhesus monkey, was one of the first American scientists to study primate social behavior.

optimistic and determined. The prospect is alluring, for, as never before, and in a measure beyond my hopes, it promises the fulfillment of my persistent dream for the progress of comparative psychobiology and the enhancement of its values to mankind through the wise utilization of anthropoid apes and other primates as subjects of experimental inquiry (P. 406).

REFERENCES

Bermant, G. and Lindburg, D. *Primate Utilization and Conservation.* New York: Wiley, 1975.

Bingham, H. Gorillas in native habitat. *Carnegie Inst. Washington Publ.* **426**: 1–66 (1932).

Bowlby, J. *Attachment and Loss,* Vol. 1. *Attachment.* New York: Basic Books, 1968.

Breland, K. and Breland, M. *Animal Behavior.* New York: Macmillan, 1966.

Brown, A. E. Grief in the chimpanzee. *Am. Nat.* **13**: 173–175 (1879).

Carmichael, L. Some historical roots of present-day animal psychology. In B. Wolman (Ed.) *Historical Roots of Contemporary Psychology.* New York: Harper & Row, 1968, pp. 47–76.

Carpenter, C. R. A field study of the behavior and social relations of Howling monkeys (*Alouatta paliatta*). *Com. Psychol. Monogr.* **10**: 1–168 (1934).

Carpenter, C. R. *Naturalistic Behavior of Nonhuman Primates.* State College: Pennsylvania State University Press, 1964.

Cuppy, W. *How to Tell Your Friends from the Apes.* New York: Liveright, 1931.

Darwin, C. *Expression of the Emotions in Man and Animals.* London: Murray, 1872.

De Haan, J. E. Bierens. Werkzeuggebtauch und Werzeugherstellung bei einen niederen affen (*Cebus hypoleucus*, Humb.). *Z. Vergl. Physiol.* **13**: 639 (1931).

Foley, J. P. First year development of a rhesus monkey (*Macaca mulatta*) reared in isolation. *J. Genet. Psychol.* **45**: 39–105 (1934).

Foley, J. P. Judgement of facial expression in the chimpanzee. *J. Soc. Psychol.* **6**: 31–67 (1935).

Forbes, H. O. *Monkeys: A Handbook to the Primates.* London: Elevan Lloyd Ltd., 1896.

Fouts, R. S. Acquisition and testing of gestural signs in four young chimpanzees. *Science* **180**: 978–980 (1973).

Gardner, R. A. and Gardner, B. T. Teaching sign language to a chimpanzee. *Science* **165**: 664–672 (1969).

Goodall, J. The behaviour of free-living chimpanzees in the Gombe Stream Reserve. *Anim. Behav. Monogr.* **1**: 161–311 (1968).

Gruber, H. E. and Barrett, P. H. *Darwin on Man and Darwin's Early Unpublished Notebooks.* New York: Dutton, 1974.

Gulliver, G. Letter. *Nature* **8**: 103 (1873).

Hahn, E. *On the Side of the Apes.* New York: Crowell, 1971.

Hamilton, G. V. A study of sexual tendencies in monkeys and baboons. *J. Anim. Behav.* **4**: 295–318 (1914).

Hamilton, W. I., Buskirk, R. E., and Buskirk, W. H. Defensive stoning of baboons. *Nature* **256**: 488–489 (1975).

Harlow, H. F. The formation of learning set. *Psychol. Rev.* **56**: 51–56 (1949).

Harlow, H. F. Primate learning. In C. P. Stone (Ed.) *Comparative Psychology.* Englewood Cliffs, N.J.: Prentice-Hall, 1951, pp. 183–238.

Harlow, H. F., Uehling, H., and Maslow, A. H. Comparative behavior of primates. I. Delayed reaction tests on primates from the lemur to the orangutan. *J. Comp. Psychol.* **13**: 313 (1932).

Kawai, M. Newly acquired precultural behavior of the natural troop of Japanese monkeys on Koshima Island. *Primates* **6**: 1–30 (1965).

Kinnaman, A. J. Mental life of two *Macacus rhesus* monkeys in captivity. *Am. J. Psychol.* **13**: 98–148 (1902).

Klüver, H. *Behavior Mechanisms in Monkeys.* Chicago: University of Chicago Press, 1933.

Köhler, W. (transl. by E. Winter) The Mentality of Apes. New York: Harcourt, Brace & Co., 1925.

Kohts, N. Recherches sur l'intelligence du chimpanzé par la méthode de "choix d'après modèle." *J. Psychol.* **25**: 255 (1928).

Kohts, N. Les aptitudes motrices adaptives du singe inférieur. *J. Psychol.* **27**: 412 (1930).

Lashley, K. S. and Watson, J. B. Notes on the development of a young monkey. *J. Anim. Behav.* **3**: 114–139 (1913).

McDougall, K. D. and McDougall, W. Insight and foresight in various animals—monkey, racoon, rat, and wasp. *J. Comp. Psychol.* **11**: 237–273 (1931).

Maslow, A. H. The role of dominance in the social and sexual behavior of infrahuman primates: I. Observations at Vilas Park Zoo. *J. Genet. Psychol.* **48**: 261–277 (1935).

Maslow, A. H. The role of dominance in the social and sexual behavior of infrahuman primates: III. A theory of sexual behavior of infra-human primates. *J. Genet. Psychol.* **48**: 310–338 (1935).

Maslow, A. H. The role of dominance in the social and sexual behavior of infrahuman primates: IV. The determination of hierarchy in pairs and in a group. *J. Genet. Psychol.* **49**: 161–198 (1936).

Maslow, A. H. Dominance-quality and social behavior in infra-human primates. *J. Soc. Psychol.* **11**: 313–324 (1940).

Maslow, A. H. and Flanzbaum, S. The role of dominance in the social and sexual behavior of infra-human primates: II. The experimental determination of the dominance behavior syndrome. *J. Genet. Psychol.* **48**: 278–309 (1936).

Maslow, A. H. and Harlow, H. F. Comparative behavior of primates. II. Delayed reaction tests at the Bronx Park Zoo. *J. Comp. Psychol.* **14**: 97 (1932).

Mitchell, G. Abnormal behavior in primates. In L. A. Rosenblum (Ed.) *Primate Behavior: Developments in Field and Laboratory Research.* New York: Academic Press, 1970, pp. 195–249.

Mivart, St. George. *Man and Apes.* New York: D. Appleton & Co., 1874.

Morgan, C. L. *Animal Life and Intelligence.* London: Arnold, 1891.

Morgan, C. L. *The Animal Mind.* London: E. Arnold, 1929.

Nissen, H. W. A field study of the chimpanzee: Observations of chimpanzee behavior and environment in Western French Guinea. *Comp. Psychol. Monogr.* **8**: 1–105 (1931).

Romanes, G. J. On the mental faculties of the bald chimpanzee (*Anthropopithecus clavus*). *Proc. Zool. Soc. London* : 316–321 (1889).

Scollay, P. Mother-infant separation in rhesus monkeys. Ph.D. dissertation, University of California, Davis, 1970.

Seay, B. and Harlow, H. F. Maternal separation in the rhesus monkey. *J. Nerv. Ment. Dis.* **140**: 434–441 (1965).

Southwick, C. H. (Ed.) *Primate Social Behavior.* Princeton, N.J.: Van Nostrand, 1963.

Strum, S. Primate predation: Interim report on the development of a tradition in a troop of olive baboons. *Science* **187**: 755–757 (1975).

Thorndike, E. L. Animal intelligence. *Psychol. Rev. Monogr. Suppl.* **2**: 1109 (1898).

Thorndike, E. L. The mental life of monkeys. *Psychol. Rev. Monogr. Suppl.* **3**: 1–57 (1901).

Tinklepaugh, O. L. Fur-picking in monkeys as an act of adornment. *J. Mammal.* **12**: 430–431 (1931).

Wakefield, P. *Instinct Display, in a Collection of Well-Substantiated Facts, Exemplifying the Extraordinary Sagacity of the Animal Creation.* London: Darton, Harvey, and Darton, 1811.

Wallace, A. R. *The Malay Archipelago.* New York: Harper & Bros., 1869.

Warden, C. J. The historical development of comparative psychology. *Psychol. Rev.* **34**: 57–85, 135–168 (1927).

Watson, J. B. Imitation in monkeys. *Psychol. Bul.* **5**: 169–178 (1908).

Yerkes, R. M. Maternal instinct in a monkey. *J. Anim. Behav.* **5**: 403–405 (1915).

Yerkes, R. M. Provision for the study of monkeys and apes. *Science* **43**: 231–234 (1916a).

Yerkes, R. M. Ideational behavior of monkeys and apes. *Proc. Nat. Acad. Sci., Washington* 2: 639–642 (1916b).

Yerkes, R. M. The mental life of monkeys and apes: A study of ideational behavior. *Behav. Monogr.* **3**: 1–145 (1916c).

Yerkes, R. M. *Almost Human.* New York: Century, 1925.

Yerkes, R. M. Robert Mearns Yerkes: Psychobiologist. In Murchison, C. (Ed.) *A History of Psychology in Autobiography*, Vol. 2. Worcester, Massachusetts: Clark University Press, 1932, pp. 381–407.

Yerkes, R. M. Genetic aspects of grooming, a socially important primate behavior pattern. *J. Soc. Psychol.* **4**: 3–25 (1933).

Yerkes, R. M. *Chimpanzees: A Laboratory Colony.* New Haven: Yale University Press, 1943.

Yerkes, R. M. and Yerkes, A. W. *The Great Apes.* New Haven: Yale University Press, 1929.

Zuckerman, S. *The Social Life of Monkeys and Apes.* London: Kegan, Paul, French and Trübres, 1933.

3
Development of Social Attachment Potential in Captive Rhesus Monkeys*

G. Mitchell

Department of Psychology
University of California
Davis, Calif.

Terry L. Maple

Department of Psychology, Georgia Institute of Technology, and
Yerkes Regional Primate Research Center, Emory University
Atlanta, Ga.

and

J. Erwin

Department of Psychology
Humboldt State University
Arcata, California

INTRODUCTION

Kuo (1967) repeatedly stressed the great complexity and variability of behavior in animals. Behavior, he said, is determined by a dynamic interrelationship of the morphological, the physiological, the developmental history, the stimulus or stimulating objects, and the environmental context. Behavioral complexity and variability have made the two concepts, instinct and learning, obsolete. Every neonate,

*This research was supported by USPHS/NIH grants MH17425, MH19760, and MH22253 to Dr. G. Mitchell, by HD04335 to Dr. Loring F. Chapman, Department of Behavioral Biology, School of Medicine, University of California, Davis, and by RR00169 to the California Primate Research Center. The authors thank the following individuals for substantial contributions to this research program: Edna Brandt, Donald Lindburg, Cheryl Stevens, William Redican, Jody Gomber, and Patricia Scollay. Patricia Jones of Davis certainly deserves some credit, having extended the same to us on occasion. An earlier form of this chapter was published in the *JSAS Catologue of Selected Documents in Psychology*, 1976, 6, Ms. 1177. Those portions reproduced here without change are copyright (1976) by The American Psychological Association and are reprinted by permission.

indeed, every embryo, has an extremely wide range of behavioral potential, only a small part of which can be actualized in a lifetime. Kuo believed, and we agree with him, that one of the principal objectives of behavioral research is exploration of behavioral potentials.

Kummer's (1971) views on behavior potential refer principally to the area of primate behavioral research:

> . . . there is a need for increased experimental research, both in the field and in laboratories, on the range of modification in response to varied environments. It is not enough to describe one variant of a species' social organization that occurs under "natural" conditions. We should investigate the modification potential of a species to its very limits, that is, to the point where the changes induced by the environment are no longer adaptive and homeostatic but lead to breakdown. I can hardly imagine a more urgent research task than to gather such knowledge about man. Insights into the tolerance limits of primates could help us in defining our own. (P. 154)

We have accepted the challenges of Kuo and Kummer by testing the behavior potential of *Macaca mulatta*, the rhesus monkey, especially with regard to the development of socio-emotional attachments between members of dyads. We have explored attachments between subjects varying in sex, age, and rearing conditions using responses to separation and reunion as indicators of attachment strength and quality.

Our methods of observing rhesus monkeys have included checklist scoring systems, clocks and counters to measure durations and frequencies of behaviors, Esterline-angus recorders for sequence, motion pictures, videotapes, Butler boxes for choice tests, and other techniques. Included in our research program have been experiments on social isolation, male care, birth, birth order, maternal experience, sex differences, attachment to nonsocial objects, facial expressions, vocalizations, "looking" behavior, interspecies attachment, and separation. It has been our opinion for some time that observational research on many different *dyads*, the members of which vary as to age, sex, rearing, species, and other factors, could provide important developmental information on normal and abnormal social and emotional potential. Specific examination of different dyads in the same laboratory, by the same experimenters, using the same methodology across studies, has insured data of a reasonably comparable nature. We have been particularly interested in developmental changes. Thus, we have oriented toward analyzing the respective roles of maturation and social experience. In so doing, we have accumulated both cross-sectional and longitudinal developmental empirical data.

HISTORICAL BACKGROUND

Research on attachment potential in the rhesus monkey and its relevance to humans relies heavily on historical research involving many species, both primate and nonprimate. In this section we would like to list and acknowledge the research which brought us to our present pursuit.

Nonprimates

With regard to nonprimates, Darwin (1872) brought the basic biological foundation of expression of emotions in man to the attention of the scientific community. A year later, Spalding's (1873) original descriptions of following in precocial birds set the stage for the role of early experience in the development of social and emotional behavior. These two publications on the biological determinants of emotion and social behavior have become groundbreaking classics in retrospect. The work on cross-species attachment potential by Kuo (1960abc), refined studies of imprinting by Lorenz (1937), and Sluckin's (1965) history of imprinting developed out of the Darwin and Spalding efforts. Scott's (1962) work on primary socialization and separation in dogs and sheep helped to bring the study of early attachment processes into the mammalian realm. Bowlby (1960), Bronson (1968), Cairns (1966), Casler (1961), and others applied these early findings to human early experience. Meanwhile, in rats, stimulation was found to play a vital role in developing emotionality (Denenberg, 1967; Levine, 1971). Theories concerning the importance of early stimulation and later social preferences, and in particular, the rate at which young organisms are stimulated, followed (Dember, 1965; Sackett, 1965). The integrative work of Cairns (1972), Gewirtz (1972), Yarrow (1972), and others on attachment, age, taxonomic position, behavioral sequences, and the contexts of *dyadic* interaction were of direct relevance to our own research.

Attachment, and especially early attachment, was the subject of several major studies dealing with nonprimate mammals (Rheingold, 1963; Altmann, 1963). Harper (1970) reviewed the ontogenetic and phylogenetic functions of the parent-offspring relationship in mammals, and Kahmann and Frisch (1952) described the relations between mothers and infants in small mammals. Both of these reviews bolstered our belief in the comparative approach to studies of social and emotional behavior. For material on attachment in birds, the reader may refer to Bateson (1969), Coulson (1966), Gareffa (1969), King (1966), Lack (1940), LeResch and Sladen (1970), Moltz (1963), Morris and Erickson (1971), Orians (1969), Salzen (1970), Selander (1965), and Stettner et al. (1971). Werner (1957) and Schneirla (1959) proposed comparative and developmental theoretical points of view based, in large part, on nonprimate organisms. Finally, Buckley (1968) edited an interesting set of readings on a systems approach to social attachment.

Humans

By bracketing early-attachment data on nonhuman primates with reviews, data, and theories based on nonprimates and on humans, attachment research gained greater applicability than it would have had on the basis of, for example, rhesus monkey data alone. Studies of human early attachment most rightly began with Freud (1933), but a more recent authority on separation and attachment is John Bowlby (1969, 1973). Bowlby's two volumes on *Attachment and Loss* are central to our area of research as are Spitz' (1949) studies of the debilitating effects of institutionalization. Wolff (1959) emphasized the important early role of smiling, bringing facial expression in humans into the research area on social attachment. Facial expressions are also of importance in nonhuman primate communication and attachment.

Nancy Bayley (1935) writing on early motor, emotional, and social development further contributed to interest in the nonverbal basis of emotion and social attachment. Dennis (1960), Goldfarb (1945), and Rheingold and Bayley (1959) added to the knowledge concerning the role of early stimulation of the human infant, especially by the mother. Fantz (1963) concentrated on the visual response of the infant to social stimuli, and Freedman (1961) and Schaffer (1963) studied fear of strangers following attachment to the familiar. Skeels (1966) convincingly demonstrated the long-term effects of deprivation of early stimulation, particularly deprivation of social attachments in institutionalized children. Skeels followed his subjects into adulthood and learned that the effects of early personal-social deprivation are tragic for humans.

Nonhuman Primates

In the area of nonhuman primatology, the early works of Carpenter (1942), Yerkes and Elder (1936), and Zuckerman (1932) set the stage for the Harlows' Wisconsin studies. Robert Hinde and his co-workers also contributed significantly to the data on early attachment and separation. Rowell and Hinde (1962) were the first to provide us with catalogues of posture, facial expression, and vocalizations in the rhesus monkey. We found these catalogues to be very useful in our research. Engel (1962), Hansen (1966), Jensen and Tolman (1962), Rosenblum and Kaufman (1968), and Seay et al. (1962) have also researched attachment and separation in macaques. Mason (1971) theorized concerning the effects of general levels of arousal and demonstrated that attachments also occur later in life (Mason and Kenney, 1974). Sackett (1965) contemplated the complexity level of the environment and of the individual macaque's early life.

In experiments with far-reaching implications, Gallup and McClure (1971) related self-recognition and recognition of conspecifics to exposure in early life. In order to know the self, an ape must know others of his own species early in life.

The behavior of adult rhesus monkeys was researched and/or reviewed by Agar and Mitchell (1975), Michael and Zumpe (1971), and Rowell (1972). The social and emotional behaviors of the wild rhesus monkey were examined most thoroughly by Lindburg (1971).

THE DYAD-SEPARATION PARADIGM

The rationale for research on attachment and separation in animals was discussed by McKinney et al. (1972). However, these authors restricted themselves to models for human depression, whereas we have a more general interest in the long-term lifetime development of all attachment capabilities. In addition, they did not use dyads but instead employed groups of four. Their methodology varied from study to study and from age to age, probably because of their interest in depression per se rather than in lifetime development of attachment potential.

We studied the attachment potential of rhesus monkeys for seven years during which time several lines of research developed. Our first means of determining the nature and/or intensity of emotional bonds was to select two individuals and observe their behaviors when they were together. Child psychologists have also done this with the human infant in studies of such things as smiling, visual following, etc. Our second method was to break the bond and to observe the results. A description of what occurs when emotional bonds are broken follows.

The Separation Syndrome

When two individuals are separated from one another, there is usually an emotional protest. This protest has been attributed (by John Bowlby and others) to feelings of emotional loss on the part of the individuals protesting. Following this protest, if the separation lasts long enough, emotional depression frequently develops. As mentioned before, this depression has been the subject of much research on both human and nonhuman individuals (cf. McKinney et al., 1972).

In our own research, we have not been as concerned with the depression stage of the separation syndrome as have others. Our desire has been to bisect, not to destroy, the dyadic bond. Consequently, most of our separations have been of brief duration. When two separated individuals are returned to one another following their period apart, reunion behaviors have told us about the nature and intensity of the bond.

Use of the Dyad

In our research, we assumed that the study of social-emotional attachment in dyads would reveal the general capability or inclination of a given individual to become attached to others; and, in addition, this procedure might reveal differential ways in which the individual becomes attached as compared to others.

We have been studying social-emotional attachment potential (made up, admittedly, of a complex assortment of index responses). These indices of attachment potential are easier to observe and measure when they are changing, when they are growing, or when they are interrupted than when static. This is a principle upon which "critical period" hypotheses are based and is not an assumption seriously questioned by most researchers.

In the dyad-separation paradigm, behaviors change rapidly. While we have observed dyadic separations in an "artificial" context, it is obvious that dyadic separations occur every day in the life of a free-ranging individual. Many of these "natural" separations are imposed upon individual dyads by factors which are extraneous to the dyad. In our research, *we* have been the extraneous factor. Yet, the emotions we are studying are natural. And so are the emotions in triadic bonds, and bonds in larger groups.

Social and emotional behaviors in a dyad, however, are much easier to measure than they are in larger groups. In addition, most of the relevant social behaviors appear in the dyad. Although there are many elements of social and emotional behavior which cannot be measured in the dyad, most of them can be. As Cairns (1972) has pointed out:

> A key to understanding developmental and contextual continuity in social behavior is the specification of how dyadic patterns are generalized and modified. (P. 73)

The different roles that have been attributed to unwieldy hypothetical constructs such as attachment can be better defined by smaller constructs which are more easily measurable by specific index responses. These index responses, however, vary with the behavioral repertoire of given dyads. Critical observational research on many different dyads, the members of which vary as to age, sex, rearing, and other factors, should provide important developmental information on the nature of normal and abnormal nonhuman and human social and emotional potential.

In nature, our subject species (*Macaca mulatta*) lives in troops. We are aware that social context determines, to a large degree, the kinds of social behaviors displayed (Anderson and Mason, 1974). Nevertheless, confident of our coverage of most of the important social and emotional behaviors, we have chosen rigorous control of experimental measurement in our quest to test attachment potential in order to avoid the multitude of complex extraneous influences present in wild troops.

Overview

The results of the work done in our laboratory from 1967 through 1974 can be categorized into three separate areas: 1) basic studies on social-emotional communication utilized in attachment; 2) nature of attachment in different dyads; and 3) separation and reunion studies in different dyads. In the first category, facial expressions, vocalizations, movements, postures, and looking behavior were studied in detail. Slides, motion pictures, and Butler boxes were employed in these studies. With regard to our studies of dyadic attachment, the intensity and form of an attachment was found to depend upon the rearing experience (e.g., isolation-reared, inexperienced mother, etc.), the age, the length of time together, and the sex of the two individuals involved. Most of our studies of this kind involved the pairing of two animals for various periods of time. In our separation and reunion studies, most of the above dyadic bonds were disrupted to determine the ways in which the individuals were different or alike in their response to this stress. We found responses to separation to be dependent upon many of the factors studied (e.g., age, sex, rearing, length of pairing, heterosexual versus homosexual pairs, species, etc.).

BASIC STUDIES OF COMMUNICATION

Initially, we completed several studies which were designed to better define the role of rhesus monkey social signals in attachment. In one of these studies, four newborn infant rhesus monkeys were raised in social isolation for the first seven months of life, and their facial expressions, vocalizations, postures, and movements compared to those of mother-reared controls (Baysinger et al., 1972). The following conclusions were derived from this study.

1) Infant rhesus raised in laboratory cages do not display facial expressions very frequently, whether they are isolate-reared or reared alone with a mother. (A search of the literature following this study indicated that facial expressions do not occur frequently in the early months of life even when peers are provided.)

2) When given the opportunity, an isolate-reared infant will spend more time looking at human observers, at the nonsocial environment, and at itself than will a control-reared infant. This is obviously because the control infant spends time looking at its mother.

3) Abnormalities in movement and self-directed behavior do not appear to develop simply because an animal has not practiced social signals.

Since the control infants in the study above spent a considerable amount of time looking at their mothers, we decided to examine the role of looking (at other animals) in the development of attachments. Our first experiments on "looking" involved the use of colored slides and a Butler box in which preadolescent ($2\frac{1}{2}$-year-old) female rhesus monkeys could press a bar to turn on or turn off a slide. The slides, in turn, were pictures of facial expressions preadolescent female monkeys preferred to look at. Lever presses for fear grimaces and threat expressions were less frequent than presses for lipsmacks and plain faces (Redican et al., 1971).

Following these initial experiments, frequencies, durations, and average durations of looking behavior were monitored for all ages and both sexes in our laboratory. We correlated the occurrence of different facial expressions with different durations per look. Lipsmacks were accompanied by looks of longest duration and seemed to be related to friendliness, affection, or interest. Fear grimaces were accompanied by quick glances at the threatening object and seemed to be related to withdrawal, flight, or submission. Threats were accompanied by looks of intermediate duration and had a stare component, with minimal eye movement, and tension around the eyes. These last expressions were related to hostility, aggression, attack, or dominance (cf. Mitchell, 1972). Characteristic facial expressions are shown in Plates 1 and 2.

It is interesting that these three emotional-motivational states appear to develop in a chronological sequence from affection to fear to aggression (Rowell, 1963). At least, the looks and expressions accompanying the states apparently develop in this order. We will return to this subject of looking behavior later in this report. First, however, we will describe the various forms of rhesus communication.

Vocalizations

Erwin (1975) summarized the literature on rhesus vocalizations and noted that individual variability and complexity were inherent in the vocal repertoire of *Macaca mulatta.* The earliest descriptions of rhesus vocalizations published by Rowell and Hinde (1962) have been employed in field studies (Lindburg, 1971; Neville, 1966) as well as in laboratory research (cf. Seay and Harlow, 1965). Rowell and Hinde separated rhesus vocalizations into two classes: harsh noises and clear calls. Harsh noises included barks, long growls, girns, and explosive coughs. The sounds which are emitted by rhesus monkeys during agonistic encounters usually belong to the harsh noise category. The shriller the noise, the more the fear, and the less the anger.

Clear calls vary in quality and frequency from individual to individual and we, like Rowell, have noted that they seldom occur in agonistic contexts (Erwin, 1975). They often occur when an animal is separated from its group in the wild, hence they are of crucial importance to us in our separation studies. However, almost any change in stimulation in the laboratory tends to evoke clear calling from rhesus monkeys.

Our own studies of vocalizations impressed upon us an appreciation for the roles of age, sex, and rearing in attachment processes involving vocalizations. A study completed in our laboratory revealed, for example, that distress calls for social contact (coo vocalizations) decreased with age and were more frequent in females than in males. At around puberty (which is earlier in females than in males), there was a relatively precipitous decrease in these calls for both sexes but particularly for males (Erwin and Mitchell, 1973).

Age and Vocalization. Rhesus infants vocalize at birth, and we saw some neonates do so before their bodies completely emerged from the birth canal (Brandt and Mitchell, 1973b). Their earliest sounds are high-pitched shrill calls, but within the second six months of life clear calls and barks occur most frequently. Our juvenile rhesus (second year of life) were extremely vocal and emitted many clear calls; however, clear calls decreased rapidly at puberty (Erwin and Mitchell, 1973).

Sex and Vocalization. In our laboratory, Scollay (1970) found sex differences in the frequency of vocalization at the early age of three months. Females vocalized more than males at every age, other than when the females had already reached puberty and the males had not (Erwin and Mitchell, 1973). Our rhesus adult females emitted more clear calls during the late follicular phase than when menstruating (Maple et al., 1974b). For other references concerning possible sex differences in macaque vocalization, see Grimm (1967) and Itani (1963).

In our laboratory, Erwin recorded the vocalizations of a rhesus monkey infant (age three months) at the time that it was separated from its mother. The recording was played to twelve young adult rhesus monkeys, six of each sex (see Erwin, 1974b). All of the adults were less active during the time when the recording was being played than before and after. The subjects also stayed near the speaker and looked at it more when the infant vocalizations were audible. Some of the females, but none of the males, responded to infant distress sounds by looking down at and grasping their nipples and by displaying retrieval or support postures typical of mother monkeys with small infants. Females also vocalized more than males in response to the recorded infant sounds.

Rearing Experience and Vocalization. Our rhesus macaques reared in social isolation usually vocalized less frequently at puberty than did our socially reared animals (Mitchell, 1968a). This suggested that either learning, excess emotionality, or both might be involved in the development of rhesus sounds. Our social isolates also cooed less frequently and out of context during infancy (Brandt et al., 1972). Infants of our young and inexperienced mothers girned, cooed, and screeched more than did our infants of experienced mothers (Mitchell et al., 1966b; Stevens and Mitchell, 1972). Even the younger, inexperienced mothers themselves coo vocalized more than did their experienced counterparts (Scollay, 1970). These findings and the report that found rhesus monkeys can learn to emit and inhibit calls for food reward (Sutton et al., 1973) showed that rhesus vocal responses were not simply emotionally based sounds, but that they were under some degree of volitional control.

Interspecies Vocal Communication. In our laboratory, Maple and Lawson (1975) paired an adult female olive baboon (*Papio anubis*) with an adult male rhesus monkey (*Macaca mulatta*). During the first thirty days of their pairing, interspecies communication occurred in the form of reciprocal vocalizations. The two animals evidently understood the meaning of the other's sounds, probably, in part, because they were accompanied by facial expressions. Since lipsmacks are very similar in the two species, this suggested that visual communication was also of importance to macaque and baboon attachment, especially to adult attachment.

Summary on Vocalization. The coo vocalization is a clear call which generally occurs in the rhesus monkey when there is a stimulus change. Separation of an infant rhesus from its mother or from a familiar environment is particularly likely to produce increases in the frequency, intensity, and quality of these calls (Seay et al., 1964). Clear calls occur less frequently in adults than in infants and less in males than in females (Erwin and Mitchell, 1973). Facial expressions to some extent replace vocalizations of this sort as monkeys mature (Brandt et al., 1972). Both infants and adult isolate-reared monkeys coo less frequently than do socially reared monkeys of similar age, and, in addition, the coo is not used in appropriate contexts by isolates (Brandt et al., 1972; Mitchell, 1968a; Mitchell et al., 1966a).

The screech is a harsh noise associated with withdrawal and/or fear. It is emitted more frequently by infant isolates than by infant controls, but this difference disappears as the animals mature (Fittinghoff et al., 1974; Mitchell, 1968; Mitchell and Redican, 1972; Mitchell et al., 1966a).

Other studies in our laboratory which suggest that early experience affects vocalization are included in our studies involving social isolation, maternal experience, maternal age, birth, and birth order (cf. Brandt and Mitchell, 1971, 1973b; Mitchell and Schroers, 1973; Stevens and Mitchell, 1972).

Movement and Posture

Movements and postures are conspicuous, hence they have the potential of being used in attachment processes. In the first month of rearing, socially reared monkeys do more climbing and jumping than do isolate-reared monkeys, whereas isolates are more inclined to walk. In addition, the movements of isolates are somewhat slower, awkward, and rigid. These differences are maintained throughout life but decline somewhat as the animals pass puberty. Isolates, however, remain awkward in their movements even beyond twelve or thirteen years of age (Baysinger et al., 1972; Fittinghoff et al., 1974). These differences in movement and posture convey information to human observers concerning the monkey's confidence and/or degree of relaxation. They convey similar information to conspecifics.

During the first two years of life, isolates display more crouching and cowering than do socially reared animals. As the isolates near puberty, crouching and cowering spontaneously decrease. By the time the isolate-reared monkeys are adults, they are no longer significantly different from controls in these behaviors. Partly because of these changes in isolate posture, young isolates are usually subordinate when compared to young socially reared monkeys, whereas the relative dominance of isolates and controls past puberty is not as easy to predict (Rowland, 1964; Mitchell, 1968a).

Isolate-reared rhesus at all ages show both qualitative and quantitative abnormalities in sexual posturing. This is particularly true of males, but it is also true of females (see also Mason, 1963a). In spite of attempts at social therapy, abnormal sexual behavior is extremely resistant to change. We will discuss our attempts along these lines later in this report.

Monkeys reared in total or partial isolation display stereotyped movements more frequently and longer than do socially reared monkeys (Berkson, 1967; Mitchell et al., 1966a). The deprivation stereotypies, such as rocking, wane with age so that by the time the isolate is two or three years of age, he usually no longer rocks. Rocking is replaced by stereotyped movements which are related to cage size, such as repeated jumping, pacing, or somersaulting (Cross and Harlow, 1965; Mitchell, 1968a). Mason (1968) has shown that isolates reared on moving inanimate surrogates do not display rocking and, in addition, show less arousal and/or fearfulness than do isolates not reared on moving surrogates.

A wide assortment of bizarre postures and movements appear in our isolates within the first month of life (Baysinger et al., 1972) and some of these apparently remain throughout life. The number of bizarre postures and movements displayed by an individual decreases as the animal matures. When the isolate-reared monkey is a fully mature adult, he usually displays only two or three idiosyncratic bizarre movements (Fittinghoff et al., 1974). These, like abnormal sexual behaviors, are extremely resistant to change.

All of the abnormalities in posture and movements discussed above interfere with effective social and emotional communication, as do abnormalities in facial expressions. Many abnormal and stereotyped behavior patterns also occur in laboratory-reared rhesus monkeys even though social experience is provided (Erwin et al., 1973).

Facial Expressions

Juvenile and adult isolate-reared animals do not use facial communication adequately. In addition, normal animals appear to be disturbed and/or confused by the looking behavior and facial expressions of animals which have been deprived of social attachments early in life (cf. Brandt et al., 1971; Brandt et al., 1972; Fittinghoff et al., 1974; and Redican, 1975).

Mason (1963ab) and Miller et al. (1967) have also shown that isolate-reared monkeys are defective in both sending and receiving facial expressions. The three facial expressions mentioned earlier appear to be particularly important in the genesis of social behavior: the lipsmack, the fear grimace, and the threat. Unfortunately, little is known concerning the development of lipsmacking in isolate-reared animals. The lipsmack, the first facial expression to develop ontogenetically has substantial value in promoting peaceful interaction between animals. Erwin has recently noted an apparent tendency for rhesus infants as young as three-months old to respond in kind to lipsmacking by a human observer. The fear grimace occurs more frequently in young (under two years) isolates than in young controls, but this rearing difference decreases markedly at or near the onset of puberty (Mitchell, 1968a; Mitchell et al., 1966a). The stare threat and open mouth threat, often accompanied by furrowed brows and head bobbing, occur less frequently in young isolates than in young controls, but this difference also disappears at or before puberty and reverses following puberty (Mitchell, 1968a).

The most frequently used mode of communication among rhesus monkeys, and perhaps among all terrestrial nonhuman primates, is looking (Mitchell, 1972). As we have stated previously, the duration of each look appears to be related to three basic emotional states. Long, wide-eyed expressionless looks appear earliest in life and are associated with affection, interest, or curiosity. They are most evident in neonates and they seem to be related to lipsmack facial expressions and to girning vocalizations. Short, quick glances appear next in the first year and seem to be related to fear grimaces and screech vocalizations. In looks of intermediate duration, the areas around the eyes look tense and a minimum amount of eye movement is displayed. These looks are related to threats and barks and generally develop after looks of affection and looks of fear (Mitchell, 1972; Rowell, 1963).

Infant monkeys display few facial expressions, but their looking behaviors are extremely important. Isolate infants display greater frequencies of quick glances than do controls. Adult isolates display a greater average duration per look than do control adults. These looks are apparently interpreted as stare threats by normal adults, but it is possible that they may be examples of fixation at the level of infantile wide-eyed looks (Mitchell, 1972).

Self-Directed Behavior

Isolate-reared infants and adults direct significantly more behavior of all kinds toward themselves than do socially reared monkeys. These self-directed behaviors apparently develop through the following sequence: affection-related, fear-related, and aggression-related (Mitchell et al., 1966a). This sequence is the same for normal animals (Rowell, 1963), but the direction in which the behaviors are displayed differs. The isolates fail to develop a clear idea of self versus nonself; and, as they reach adulthood, they direct more and more hostility toward the self.

These findings are of relevance to some exciting research by Gallup and McClure (1971) alluded to in an earlier section. Isolate-reared chimpanzees do not recognize themselves in mirrors as do socially reared chimpanzees. Perhaps the precursors of self-recognition (obviously present in apes) are present in normal rhesus monkeys but are lacking in isolates. Since isolates are also hostile toward themselves (i.e., they bite themselves), it may be that the distinction between self-recognition and self-love is as difficult to make in monkeys as it is in man. The failure to perceive self could interfere with peaceful social communication, and it is probable that the monkey can come to know itself only through interaction with others. It is interesting that self-slapping and self-biting, along with bizarre movements and abnormal sexual behavior, remain in the adult isolate's behavioral repertoire even after other social behaviors such as grooming, play, and sustained peaceful social behavior have returned as a result of therapy.

Arousal and Alternatives

In closing this section on basic communication, it is probably important to consider the state of arousal of the animal and the alternatives open to the organism. There is no question that mature social isolates are often excessively aroused, and especially so in those situations where they display other-directed or self-directed aggression. There is also no question that they are relatively incapable of utilizing other social objects to decrease that arousal. The only increases in arousal they are able to tolerate are deviations which they themselves can produce and control. With no social alternatives open to control arousal, social stimulation produces intolerable levels of excitement which, in turn, accentuate the abnormalities in communication already discussed.

Even in feral-reared and/or normal adult males we have seen cases of extreme self-directed aggression result when the animal's arousal was increased and when there were apparently no arousal-reducing alternatives aside from behaviors directed toward the individual's own body (cf. Erwin et al., 1973b; Maple et al., 1974a). During the separation of an adult feral male from a seven-month-old infant which the male had reared, the normal male bit himself so severely that, had he not received immediate veterinary care, he would have bled to death. Clearly, unkind situations can be devised which make normal animals behave as do social isolates, at least in some ways. Abnormal self-directed aggression, although less likely, can and sometimes does appear in monkeys which have not experienced early social isolation. Humans who have not had histories of social deprivation may also display such abnormal aggressive responses to conditions of extreme stress in which there are no other arousal-reducing alternatives.

We do not mean to imply here that arousal or excess emotionality can account for all of the abnormalities in attachment that we have observed. Despite low levels of arousal or emotionality, isolate-reared monkeys remain incapable of some forms of effective communication. Inadequate and/or absent sexual posturing as well as bizarre movements appear even when the isolates are relaxed. There are quite obviously cognitive as well as emotional deficits (Mitchell, 1968a; Gomber, personal communication).

THE NATURE OF ATTACHMENT IN VARIOUS DYADS

Birth (Mother and Neonate)

In the wild, rhesus monkey births usually occur in the spring and at night (Bo, 1971). The female has been assumed to deliver away from the troop and alone; however, other evidence suggests (see Caine and Mitchell, this volume) that this is not so. Prior to parturition, there is a slight increase in adult heterosexual activity and aggression. These changes have been attributed, at least partly, to increases in estrogen before birth (Agar and Mitchell, 1975; Hausfater, 1972; Koford, 1966; Lindburg, 1971).

Tinklepaugh and Hartman (1930) first described night delivery, squatting, straining, genital exploration, licking (see Lindburg and Hazell, 1972), delivery of the baby, delivery of the placenta, and placenta eating in *Macaca mulatta.* These deliveries were of an average duration of two to three hours, following a gestation period of 160 to 180 days. The births usually took place at night (see also Harms, 1956; Hartman, 1928; Jacobson and Windle, 1960).

Rhesus monkey deliveries in our laboratory usually occurred at night also (Brandt and Mitchell, 1971, 1973b). Additional signs of labor seen at Davis were: restlessness, self-scratching, body shakes, rigid tail wagging, hopping on hindquarters, and stretching of the hind legs. The duration of labor ranged from about fifteen minutes to four or five hours, with the longest labor occurring in primiparous females and/or in abnormal deliveries. There were usually five to forty labor pains involving obvious straining and stretching. Occasionally, prior to birth, the amniotic sac bulged out the vagina. Many infants aided in their own deliveries by clasping the mother with their hands before their legs had emerged from the birth canal. One-fifth of the infants vocalized during birth. Mothers rarely vocalized if left alone during delivery. In the dark, infant vocalization might help the mother locate the infant should she temporarily lose sight of it following delivery.

The nature of attachment between a mother and her infant depends upon the age and/or experience of the mother (Brandt and Mitchell, 1971; Brandt and Mitchell, 1973b). For other factors affecting birth and emotional attachment, see Koford et al. (1966), Meier (1964, 1965), and Windle (1967). We have noted that primiparous females have longer and more difficult deliveries than do multipara, and that primipara are usually awkward and anxious during and following delivery. If the primiparous female has been reared in social isolation, she may be indifferent or brutal toward her first born (see Mitchell and Brandt, 1975; Mitchell and Schroers, 1973) but not toward her subsequent infants.

There is no prolonged rest period for the mother following parturition, and though she appears to be tired, she does not seem extremely weakened by the birth. If a second female near term is present during delivery, she may be facilitated into labor by the sight or smell of the first delivery. Other females show interest in the birth such as watching intently, lipsmacking, arching their backs, touching their own vaginas, and investigating the birth fluids (Rowell et al., 1964). An attending female sometimes grooms the mother in labor and threatens human observers. At times an attending female will attempt to kidnap the newborn, and occasionally she succeeds.

If an adult male is present, he appears to be mildly anxious yet displays intense curiosity during delivery (Mitchell and Brandt, 1975). He watches closely, approaches, and grooms the mother, paying special attention to her posterior. He occasionally becomes sexually aroused and may try to mount the female. He threatens observers, sometimes intensely. Following delivery, he approaches the female to get a better look at the infant and/or placenta. He touches and manipulates the placenta and may lick his fingers but he does not attempt to eat the placenta. He tries to touch the infant and may try to take it from the mother (Redican and Mitchell, 1973). At these times, the infant is in some danger of injury by an adult male's over-eager attempts to inspect it or take it from the mother. His overall manner, however, is not one of aggressiveness but is rather one of interest or intense curiosity. The female usually tries to withdraw from the male's approaches. She sometimes fear grimaces. Females are not always pestered by males, however. On one occasion, a male was groomed by the female one hour after the delivery as he lay near her (Mitchell and Brandt, 1975). For more detailed reports on rhesus birth, see Brandt and Mitchell (1971, 1973b) and Mitchell and Brandt (1975).

Mother and Infant

The nature of the attachment between a mother and her infant throughout the first year of life depends also on the age of the infant, and the sex of the infant (cf. Baysinger et al., 1972; Brandt et al., 1972; Fragaszy and Mitchell, 1974; Mitchell, 1971; Mitchell, 1977). We examined the role of the infant in early attachment by comparing infants in early isolation with infants in the presence of their mothers. The mother-reared infants were much more active and outgoing (Baysinger et al., 1972). (See Plate 4a.)

We also compared the mother's relations with the male infant to the mother's relations with the female infant in thirty-two mother-infant pairs. Mothers had more physical contact of a relaxed type with female infants. They also restrained female infants more frequently than male infants. Mothers of males, on the other hand, withdrew from, played with, and presented to their infants more often than did mothers of females. Males bit their mothers more often than did females. Ventral contact, nipple contact, and sustained embracing decreased with the age of the infant while threats and punishment increased (Mitchell, 1968b).

The above findings were corroborated and refined in two later studies (Mitchell and Stevens, 1969; Mitchell and Brandt, 1970) in which it was shown that the experience of the mother also played a role. Eight inexperienced rhesus mothers were matched with eight experienced mothers with regard to delivery date and sex of infant. The inexperienced mothers looked at, threatened, fear grimaced, and lipsmacked toward strange mother-infant pairs and toward human observers significantly more frequently than did the experienced mothers. In addition, the inexperienced mothers stroked or petted their infants more frequently than did the experienced mothers. Primiparous mothers thus appeared to be more anxious or concerned for their infants' welfare than did multiparous mothers. In the second three months of the infants' life, however, maternal experience waned in importance as the sex of the infant and individual differences became more important factors influencing maternal or infant behavior.

Follow-up studies of infants of experienced and inexperienced mothers at six to eleven months (Stevens and Mitchell, 1972) and in adolescence (Mitchell et al., 1966b) have found no severe behavioral disabilities in the progeny of either group of mothers. First-born infants and adolescents vocalized more frequently and displayed more stereotyped movements than did later borns, but these quantitative differences did not impair their social behavior or attachment potential.

Comparisons of the mother-infant relations of experienced and inexperienced normal mothers with the mother-infant relations of experienced and inexperienced isolate-reared mothers revealed an interesting interaction effect between early deprivation and later experience. Young, inexperienced mothers having had early socialization are protective and anxious; whereas, young inexperienced mothers reared in social isolation are brutal and/or indifferent with their first borns. With the second infant, the normal mother becomes less protective, slightly more punitive, and more relaxed; whereas, the isolate-reared mother becomes more protective, less punitive and more normal (Mitchell and Schroers, 1973). When a large sample of inexperienced and experienced mother monkeys (including all levels of early experience) is examined, there is far more variability among inexperienced than among experienced mothers (Mitchell and Schroers, 1973); thus, it seems probable that variability in attachment potential may be greater among first-born infants than later-born infants.

Infant with Infant

Eight first-born rhesus monkey infants were compared with eight later-born infants regarding attachment potential (Stevens and Mitchell, 1972). All subjects were paired at six to eleven months of age with six sex-balanced rhesus infants of similar age. First-born infants vocalized more frequently, were more active, and displayed more stereotyped behaviors than did later-born infants. The first-born infants were also more assertive and sociable and were in the proximity of their partners more frequently than were the later-born infants. Sex differences in infant rhesus monkey behavior were found which resembled those reported elsewhere. However, in our samples of first-born and second-born infants, we did not find greater behavioral variability in the first-born infants as we had expected. Our prediction of greater variability in first-born infants rested primarily on an interaction effect involving the early experience of the mother as well as the parity of the mother (Mitchell and Schroers, 1973), and we did not have isolate-reared mothers in our sample.

Juvenile-Juvenile, Preadolescent, and Adolescent Dyads

In our studies of juveniles, two individuals of the same sex spent most of their time together exploring the environment, playing vigorously (sometimes with mild hostility), mounting, presenting, and grooming. The frequency of protection, retrieval, and embracing was greatly diminished relative to mother-infant pairs (Erwin et al., 1971).

The length of time two juveniles or preadolescents had been together affected the nature of their attachment. Twelve mother-reared rhesus monkey preadolescents, six of each sex, which had never previously been allowed direct physical contact with the opposite sex, were paired at three years of age. Initial responses to the pairing ranged from near indifference to violent aggression. The most common response of the males was sexual mounting, although their mounts were at first awkward and poorly timed. The females were typically slow to perform adequate sexual presents, and in one pair a female aggressed and repeatedly mounted her male cagemate. The pairs lived together for either one, two, or three weeks, and it was apparent that even these brief periods of exposure produced progressively increased rapport and attachment within pairs. The males remained fairly stable in their behavior toward the females as the length of the pairings increased, while the females seemed to relax and become more confident. As a result, there was an increase in social contact between the members of a preadolescent heterosexual dyad as the length of pairing increased (Erwin et al., 1973a).

A comparison of rhesus having peer experience in the second year with those not having peer experience in the second year suggested to us that second-year social experience was important (though less important than the first year) for the optimal development of later sociosexual behavior (Erwin and Mitchell, 1975; Erwin et al., 1974a).

Preadolescent with Infant

In addition to providing basic information on the nature of possible attachments between preadolescents and infants, our preadolescent-infant pairs also taught us much about: 1) preparental behavior in both male and female rhesus monkeys; 2) sex differences in infant monkeys; and 3) the effects of early social isolation.

We paired eight male and eight female isolate and control infants (eight months of age) with eight preadolescent rhesus monkeys (both males and females of thirty to thirty-two months of age) (Brandt and Mitchell, 1973a). The male preadolescents were more aggressive toward the infants than were the females, who were at times very gentle and maternal. The males did show some affection for the infants, especially for the females, but this affection was awkward and often rough. The male infants elicited and emitted more aggressive behavior in their pairings with preadolescents than did the female partners. However, they also elicited and emitted more play. There were more social behaviors directed toward normally reared infants, but social interactions between preadolescents, particularly females, and the isolate infants increased significantly over time. Normally socialized female preadolescents seemed to provide some social therapy for at least two of the isolate-reared infant rhesus monkeys. However, social experience with preadolescents in late infancy did not permanently reverse the deleterious effects of early isolation (Erwin et al., 1974a). Bizarre movements and postures remained, and inadequate or absent sexual posturing was still characteristic of isolate-reared monkeys.

Adult Male with Infant

In our laboratory, Redican and Gomber demonstrated: 1) that it was possible for an adult male rhesus macaque, either feral born or isolate reared, to raise a young infant in the absence of a mother without inflicting severe trauma on the infant; 2) that infants so reared did not exhibit marked behavioral pathology; 3) that attachment appeared to increase in strength over time, particularly in the adult male; and 4) that for the male infant at least, and probably also for the female infant, the form of play exhibited in the dyad was of a high frequency and intensity not characteristic of mother-infant pairs. Future research on male care should explore the effects of the shift in "identification" models from adult female to adult male on sexual, aggressive, parental, play, and general attachment behaviors in infancy and adulthood (cf. Redican et al., 1974; Mitchell et al., 1974).

While caring for infants, three adult male isolate-reared rhesus (ages fourteen- through seventeen-years old) showed marked improvements in social interaction with infants. Early social deprivation did not permanently impair these rhesus monkeys' abilities to form effective relations with conspecifics. Moreover, the infants so reared did not imitate any of the bizarre behaviors of their abnormal adult male "parents."

As for the abnormal adult males, they often swayed tirelessly, self-clasped, self-bit, and self-slapped (the self-punishment usually resulted in the infants going directly to the males and touching them). These particular abnormal behaviors, however, were solitary behaviors. In the isolate males' social behavior toward the infant, much improvement was seen. For example, the adult male isolates played with the infants in a manner indistinguishable from that of the control males, and they groomed the infants intensely and in response to social cues. In addition to responding normally to social cues, the isolates also sent appropriate signals, such as threats toward the infants when the infants disturbed them, or sent threats toward a human observer who threatened the infant. The isolate's social relations with the infants seemed to be impaired only with regard to a *complete absence of social sexual behavior* (Gomber and Mitchell, 1974). This abnormality remained despite two or more years of close living with a nonthreatening infant, and despite improvements in most, if not all, other aspects of social-emotional behavior.

Overall, we demonstrated that while adult male rhesus monkeys do not care for infants very frequently in groups which include adult females, they are certainly capable of caring for them when given the appropriate opportunities in the laboratory. Clearly, the potential for attachment in rhesus males is much greater than that which they actually show in the forests of India. Moreover, male care is not merely masculine maternal behavior; it differs in form and frequency. (See Plate 3.)

While we were not quite sure what would happen to the infants that were reared by these adult males (controls or isolates), we would not have been surprised to find changes in their psychosexual behavior as adults and, perhaps, in their later care-giving behavior (cf. Mitchell et al., 1974). Other data gathered in our laboratory on adult males with infants appear in Brandt et al. (1971) and in Redican and Mitchell (1974).

Adult Male with Adult Female

While sexual behavior between adult monkeys has been fairly well studied both in captivity and in the wild, the nature of the emotional bond between adults of a heterosexual pair has not been a topic of emphasis. Field studies indicate that emotional bonds exist between adult male and adult female rhesus monkeys, but the characteristics of these bonds have not been closely studied and/or discussed in comparison with bonds in other dyads (cf. Agar and Mitchell, 1975). What potential for emotional attachment exists in adult heterosexual rhesus dyads?

We tested four heterosexual pairs of adult rhesus for six days to observe the manner in which they became attached to one another. Males rarely vocalized but often lipsmacked. Females vocalized, but not as frequently as mothers or infants in mother-infant dyads, and far less frequently than juveniles. The males rarely threatened, but the females threatened frequently, particularly shortly after the animals were put together (Maple et al., 1973a, 1974b). Most of the females' vocalizations and threats were apparently effective in inciting the male to threaten intruders and in exciting the male to sexually mount. Most of the information we obtained concerning attachment potential in these particular dyads was gathered when two adults were separated from one another. We will therefore save further discussion of this for the section on separation.

Adult Male with Adult Male

We also collected some limited data on male-male pairings, involving both isolate-reared and normally reared adult males (Fittinghoff et al., 1974). Four pairings between adult isolate and adult control males were recorded on 16-mm color film. In each pairing an isolate and a control male met in a 6 ft × 3 ft × 3 ft test cage, from which a center divider could be removed and replaced. Three of these four confrontations consisted of brief bouts of fighting interspersed with very much longer bouts during which the two males sat and threatened, yawned, ground their teeth, bluffed each other, explored the cage, and "fed" on debris. In the fourth pairing there was no fighting, apparently because the adult male isolate climbed the walls, screech vocalized, withdrew, yet threatened as well as fear grimaced.

In the three dyads where fighting was observed, most of the bouts of aggression were initiated by isolates. Two of the isolate males actually dominated their control opponents by persistence and confusion rather than by skilled fighting. Despite injuries, these isolate males repeatedly approached the control males and stared at them while attempting to "mount" themselves (self-clasping) and while masturbating. They apparently became dominant by confusing the controls with bizarre behavior (Fittinghoff, et al., 1974). Despite the irascible nature of adult male rhesus monkeys, not all relationships in the male-male dyad are fractious. In fact, there is some potential for enduring adult male homosexual attachment. (See Plate 4b.)

Interspecific Dyads

In 1973, Maple initiated studies of the development of social attachment between two different species (actually, representatives of different genera). In these studies we reached the acme of the tolerance limits in attachment for our subjects. One of our subjects was a twelve-year-old adult male rhesus monkey (*Macaca mulatta*), the second subject was a twelve-year-old female olive baboon (*Papio anubis*).

The specific methods used in the pairing are described elsewhere (Maple, 1974a).

At first, the rhesus monkey male dominated the female baboon, but within ten days the baboon became dominant and remained so for six months. From six to twelve months the bond between the two strengthened. There was increased proximity and reciprocal grooming. After almost a year together, the first copulation was observed at which time the rhesus male gained equal status with the baboon. (See Plate 4c.)

Four other baboon-macaque pairs have been studied by Maple. These were all juvenile pairings involving two same-sexed and two other-sexed dyads. Choice tests before and after intergeneric exposure have indicated that the baboon-rhesus bonds hold up even when conspecifics are provided (Maple, 1973). That is, the monkeys and baboons prefer their *contra*specific cagemates to *con*specific strangers. (See Plate 4d.)

Thus, baboons and rhesus can reverse conspecific attachments and acquire a strong social attachment to a member of an alien species. Most interesting, however, is the finding that even adults can become socially attached to animals of another species. The older the animal, however, the longer it seems to take.

Maple's studies of interspecies social behavior are important to an understanding of primate attachment for at least two reasons. First, the observation of interspecific pairs provides a setting for species comparisons that is, in some ways, more powerful than is observation of intraspecific pairs. Second, interspecific pairs test the generalizability of attachment processes in the primate order. In short, they make our studies of rhesus dyads more useful because they can be shown to be more general.

As Kummer (1971) has stated, the study of attachment *potential* is something that has been neglected in primate behavior. Like our adult males who rarely, if ever, become strongly attached to infants in the wild, yet who do so in our laboratory setting, these interspecific dyads show an amazing potential and plasticity for attachment, and this occurs within the primate order of which *Homo sapiens* is a member. Human beings are the most flexible of all primates, capable of generalizing love to the most unlikely of animals. For more detailed information on these studies, see Maple (1974ab) and Maple and Lawson (1975).

SEPARATION AND REUNION OF DYADS

Birth (Mother and Neonate)

We filmed many births and recorded the behaviors of both the mother and the neonate (Mitchell and Brandt, 1975). We know that there is a strong prewired potential for mother-infant attachment at parturition. However, we have no systematic data on the effects of separation upon the behavior of mother-neonate dyads.

Although we did not complete studies of separation and reunion for this dyad, we do have information of relevance to early mother-infant separation. In our projects on adult male care we, of course, separated infants of under thirty days of age from their mothers and gave the tiny infants to adult males. During these separations, the mothers protested violently by screeching and by directing many fear grimaces and threats toward the human intruders. The infants also screeched loudly and, after less than an hour, began cooing repeatedly. Most of the infants became depressed after a day or two. When the transfer from maternal care to male care was not made within two days, one male infant started to suck his own penis for prolonged periods of time. He recovered from this behavior after he had been paired with an adult male (Redican and Mitchell, 1973).

We also have anecdotal records of the behavior of one adult female following a breech delivery in which the infant was stillborn. This female did not display screeches or coo vocalizations as did females which had been separated from living infants. Moreover, she refused to adopt another infant two days later. Perhaps the possible relationship between difficult delivery, even Caesarian delivery, and failure to adopt (or potential to become attached) should be studied in more detail (see Brandt and Mitchell, 1971).

Mother and Infant

Mother-infant separations in monkeys have been related to early mother-infant separations in humans. The two most well-known theories of early human separation are those of Bowlby (1958) and Engel (1962). In Bowlby's theory, the response of the infant at separation is attributed to the activation of basic instinctual responses. These responses include stop and go mechanisms. Without the mother the instinctual responses of the infant do not stop and anxiety results. Engel (1962), however, suggests that these instinctual responses are physiological responses which he labels anxiety and depression withdrawal. Protest activity following separation leads eventually to a need to conserve energy. At this time, anxiety changes into depression.

The criticisms of Engel's and Bowlby's approaches center on the absence of good data at different developmental ages (cf. Casler, 1961; Heinicke and Westheimer, 1965; Spitz, 1950). This is where our monkey research comes in. Standardization is easier when monkeys are used and monkeys react to maternal separations in ways similar to human infants. In our laboratory, Scollay (1970) compared the separation and reunion responses of mother-infant dyads at varying infant ages. Prior to Scollay's study, numerous other mother-infant separation studies had involved macaque infants ranging in age from four months to thirteen months. Scollay and other co-workers at Davis extended these studies to include dyads of all ages, of both sexes, of varying rearing conditions, and even of different species.

The protest response of six- to eight-month-old infant monkeys used in separation studies has been described as rapid movement about the cage and loud cooing. After about forty-eight hours the infants become very inactive and depressed. This corresponds to Bowlby's grief phase. Only occasional reports of symptoms resembling the detachment phase had been observed in monkeys until the rather extensive study conceived and carried out by Scollay. For her study, twenty-four infant rhesus macaques were separated from their mothers for forty-eight hours, in a three-phase experiment—preseparation, separation, and postseparation. Each phase lasted two days because Scollay did not intend to produce depression in her infants (in macaques, depression usually sets in after forty-eight to seventy-two hours). During separation, the mother was moved to a similar cage in a different room. The infant remained in a two-cage observation unit. The separated infant was placed in one cage and a nonseparated mother-infant stimulus pair was placed in the other. At reunion, the mother was returned to the infant rather than the infant to the mother.

The infants were divided into three groups, each initially separated at a different age: eight weeks, fourteen weeks, and twenty weeks. Twelve of the infants were male and twelve were female (four of each sex in each age group). The two days of preseparation data were used as baseline and the separation and postseparation data were compared with them.

All infants responded to the loss of their mothers in generally the same way. They showed increases in the amount of locomotor activity. Self-play dropped out completely during separation. Although the infants exhibited stereotypies in the home cage, none did so in the observation unit during preseparation. During separation, however, stereotyped movement returned. The stereotypies consisted of slow bouncing in one place, with an undulation of the trunk in eight of the eleven animals displaying them. There was also an increase in digit sucking and scratching during separation. Infant lipsmacking and fear grimaces increased in the mother's absence, but the most striking change during separation was an increase in infant vocalization. All infants showed increases in screeching, barking, and particularly in cooing.

On the second day of separation Scollay noted that the infant spent increasing amounts of time sitting quietly, manipulating either the cage or self. Had the separation interval been extended to three days, depression would undoubtedly have developed. The only sex difference seen was related to the development of inactivity and perhaps depression. The males were more active than females prior to separation and showed slight increases immediately following separation. The females, on the other hand, were much more active than the males during separation. Furthermore, while the males definitely showed decreased activity on the second day of separation, the females' activity decreased only slightly. Thus, it appeared that the tendency toward depression was greater for male than female infants.

Age differences were striking. The fourteen-week-old infants did not fall intermediate between the eight-week-old and the twenty-week-old infants. Their activity increased during separation, and they exhibited more stereotyped behavior than did the other two groups. They were the only infants to show increased frequencies of environmentally directed behavior. The fourteen-week-old infants also cooed and screeched more, but barked less than the other age groups.

The mothers showed initial increases in activity and in cage shaking in response to separation. The frequency and duration of environmentally directed behaviors by the mothers, along with cooing and barking, increased during separation. The mothers' activity and vocalization increased the first day of separation, but decreased the second. Each mother reacted to separation in the same way, regardless of the sex or age of her infant.

In Scollay's study 75 percent of the reunions involved instant contact and embracing. Pairs typically established ventral contact promptly, but usually not nipple contact, and the infant closed its eyes almost immediately. In the other 25 percent of the reunions, the mother retrieved the infant, but the infant screeched as she did so. Ventral contact was established; but after some period of time, from one to five minutes, the infant broke contact and withdrew from the mother. Withdrawal lasted from a few seconds to more than an hour. During the longer periods of infant withdrawal, the mother made repeated attempts to retrieve her infant. Usually she was not successful, and if she was successful, the infant soon withdrew again. Often the mother chased the infant about the cage with the infant cooing and engaging in stereotyped behavior. Eventually, successful and permanent reunion occurred. Permanent retrieval was usually preceded by grooming. The mother moved progressively closer until she had established ventral contact. The infant then closed its eyes and a normal reunion sequence followed. Of the six infants (25 percent) who withdrew from their mothers, four were males and two were females. Three were eight-weeks old, two were fourteen-weeks old, and one was twenty-weeks old.

After reunion, whether immediate or delayed, the infants spent more time near and in contact with their mothers than they had prior to separation. There were increases in time spent in the same part of the cage, in ventral contact, and in frequency of nipple contact. Approaches and withdrawals decreased, indicating longer periods of continuous contact or proximity. Time spent grooming also increased, whereas maternal punishment and rejection decreased. An exception, however, was in the fourteen-week-old infants, half of whom (4) were rejected or punished by their mothers more often after the separation than they had been before separation. At reunion, Scollay noted decreases in activity for both mothers and infants. Again, however, the fourteen-week-old infants were exceptions. All infants showed less environmentally directed and self-directed behaviors following reunion. On the first day following reunion, very few infants left the mother's ventral surface, while on the second day, most infants began to show preseparation behavior patterns, venturing away from the mother for brief periods.

All infants in the Scollay project, regardless of age or sex, reacted to separation in the manner previously described in the rhesus monkey separations done by Seay and Harlow (1965). In addition, some exhibited behaviors suggestive of the third phase of Bowlby's separation syndrome. This was particularly true of the males. Their depression (grief) was also more obvious, and they had a greater tendency to withdraw from their mothers. Bowlby has related withdrawals in children to a history of repeated separation. Scollay's monkeys had been separated from their mothers for the few brief minutes each month that it took to weigh or tatoo them. Because these separations were brief and few, however, it is unlikely that the withdrawals in Scollay's studies could be attributable to these maintenance factors.

Hansen (1966) described a period during which the rhesus mother is increasingly punishing toward and rejecting of the infant. This period begins in the third month and declines in the fifth. Certainly the fourteen-week-old and twenty-week-old infants had been experiencing some rejection prior to the separation. In addition, we have found that normal mothers tend to be more rejecting with male infants than with females, particularly if the mothers are multiparous or experienced. These age and sex factors relating to maternal rejection may play a part in infant withdrawal at reunion.

Since it was the infants near three months of age (fourteen weeks), particularly the males and particularly those having multiparous mothers who seemed most sensitive to separation, it might be reasonable to suggest that increased levels of maternal rejections sensitize the infant to separation. In the normal rejection of her infant, the mother is, in effect, imposing a short-term separation. Increasing numbers of these separations may sensitize the infant to future separations (cf. Mitchell et al., 1967), and this might explain the appearance of withdrawal from the mother in some of these animals. Thus, unstable mother-infant bonds prior to separation produce greater distress at separation than do stable bonds. Erwin et al. (1973a) found the same thing to be true of unstable preadolescent-preadolescent bonds.

The period of withdrawal could also have been related to the fact that the fourteen-week-old infants exhibited less affectionate post-separation effects than did the other infants. In punishing and rejecting, the mother is prohibiting the infant from spending time in contact with her and thereby lowering the reunion effect.

Scollay separated each of her twenty-four infants from their mothers for a second forty-eight-hour period at six months of age. Both the infants and the mothers responded as in the first separation. Infant withdrawal was also observed after the second separation; this time it involved about 33 percent of the infants, again primarily males. At this age, the withdrawal seemed to be intensified somewhat.

In summary, Scollay found that the age and sex of the infant were not important factors in the mothers' reactions to separation, but that they were important factors in the infants' reactions to separation. In addition, she found that the males, more than the females, tended to become inactive on the second day of separation. The fourteen-week-old infants differed from both the older and younger infants in several respects, at least during the first separation series. The presence of withdrawal at reunion might be related to the age and sex of the infant, as well as to maternal rejection and/or to a history of prior separation.

Another facet of the Scollay separation study was the measurement of the behavior of the nonseparated neighbors (the mother-infant dyads used as stimulus animals) during the two-day separations. The mother-infant pairs that observed a separation, but were not themselves separated, displayed increased physical contact. The observing mothers, but not the clinging infants, also emitted an increased frequency of coo vocalizations. Thus, the macaque mother and infant apparently display separation symptoms to only the sight and sound of a separation. The mother-infant dyad is particularly sensitive to separation. In monkeys, observing a separation may even produce responses resembling feelings we would label "empathy" in humans.

Because the Scollay studies had indicated that the nature of the relationship prior to or at separation was related to the effects of seperation, we designed a study comparing the effects of two extremely different preseparation rearing environments on the behaviors of infant rhesus monkeys during separation (and subsequent reunion) from these rearing environments. Eight of our seven-month-old infants (four males and four females) were separated either from their isolation cages or from their mothers (Brandt et al., 1972). Comparisons between isolation-reared and mother-reared monkeys were made before, during, and after early separation from rearing environment or mother. As in the Scollay (1970) study, two-day preseparation, separation, and reunion phases were employed.

Stereotyped movements appeared in the control infants in response to separation, while isolate stereotypy occurred before and after as well as during separation. Both isolates and controls showed increased self-directed behaviors during separation. Vocalizations specifically associated with separation in the controls were used less often and outside of the separation stage by the isolates. It was concluded that social rearing prior to separation was the factor primarily responsible for the appearance of the behaviors seen at separation.

The Brandt et al. (1972) study also compared the effects of separation on mothers to the effects on infants. Mothers responded with many facial expressions and few vocalizations (distress calls). Scollay (1970) had reported a similar finding.

Juvenile-Juvenile and Preadolescent-Preadolescent Dyads

Preadolescent monkeys of the same sex were of interest in one of our separation studies (see Erwin, this volume). Preadolescent same-sexed pairs were separated and a potential Bowlby-like syndrome was researched with relevance to the mother-infant separations seen in our own laboratory and elsewhere. Differences between preadolescent and mother-infant dyads were most marked during reunion. Our preadolescent monkeys (of like sex) initially ignored or avoided each other during reunion, while mothers and infants directly returned to one another in 75 percent of the cases (Erwin et al., 1971).

We also separated and reunited heterosexual pairs of preadolescents (Erwin et al., 1973a). Twelve rhesus monkeys, six males and six females were paired across sex for one, two, or three weeks, after which each dyad was separated for two days. Those pairs that were together longest (three weeks) exhibited the greatest amount of social contact before separation and after reunion but responded to separation with the *least* apparent distress. The separation responses of the preadolescent females in these heterosexual dyads were particularly affected by the length of the pairing. It is possible that the relation between length of pairing and response to separation found here applies only to certain ages. These results, along with a similar finding reported by Scollay and discussed earlier in the paper, calls into question the use of the intensity of separation distress as an index of the initial strength of the social bond. But, even more importantly, the result emphasizes the need for further research on separation responses and their relation to various ages, to sex differences, and to varying kinds and degrees of attachment.

In a study involving responses of like-sexed adolescents to reunion after two years of separation (Erwin et al., 1974b), we confirmed the existence, specificity, and persistence of affective bonds which had been established during the second year of life (see Erwin, this volume). Twelve $4\frac{1}{2}$-year-old rhesus monkeys, six of each sex, were reunited with the like-sexed peer with whom they had spent their entire second year of life. Erwin compared the responses of these animals during reunion with their familiar peers to the responses of the same animals when paired with unfamiliar cagemates. Despite the fact that members of familiar pairs were separated from each other two years earlier, members of familiar dyads displayed less aggression, less fear/submission, less disturbance, and more affiliation than did members of unfamiliar pairs. The difference was particularly striking when familiar female dyads were compared to unfamiliar female dyads. Erwin concluded that like-sex attachments established during the second year of life could survive a two-year separation (Erwin et al., 1974b), and that female-female bonds at these ages were less resistent to extinction than male-male bonds.

In a companion study of cross-sex attachments in adolescence, Erwin and Flett (1974) established that heterosexual attachments were also enduring and specific. Twelve $5\frac{1}{2}$-year-old rhesus monkeys, six of each sex, were reunited with opposite-sexed peers with whom they had been paired for six months during early adolescence. The responses during reunion of the familiar animals, despite the fact that these animals had been separated for two years prior to the reunion, were primarily affiliative or positive. Familiar heterosexual dyads never aggressed one another. Males, however, aggressed unfamiliar females and both sexes directed more threats toward unfamiliar than toward familiar animals. Although males mounted familiar females more often than unfamiliar ones, it was apparent to Erwin and Flett, that heterosexual attractiveness was not based exclusively on familiarity (Erwin and Flett, 1974). The results of the Erwin et al. (1974b) study and the Erwin and Flett (1974) study were directly compared in Erwin et al., (1975).

In summary, attachment in the rhesus during the second, third, and fourth years of life can be enduring and quite specific to either a like- or unlike-sexed peer. The strength of some of these bonds appears at times to rival those of mother-infant bonds seen much earlier in life. Thus, there is potential for very strong social-emotional attachment in the juvenile, preadolescent, and adolescent stages of life.

Preadolescent with Infant

Each of eight eight-month-old infant rhesus monkeys was paired with a (thirty- to thirty-two-months old) preadolescent conspecific for two months (Brandt and Mitchell, 1973a) and then separated. Four of the infants were mother reared and four were isolate reared. Separation responses were compared with data from preseparation and reunion phases of the study for all pairs. The control infants responded with disturbance that was more often related to the separation than did the isolate infants. The results also indicated that: 1) although preadolescent males at times interacted with infants in a parental fashion, preadolescent females showed a greater capacity for parental-like behavior; 2) both preadolescents and normal (mother-reared) infants contributed to the development of a social bond; and 3) isolate-reared infants contributed little to the development of a social bond and were relatively less valued as social partners by their preadolescent cagemates (Maple et al., 1975). In this particular dyad, early social experience apparently predisposes an infant to actively promote peaceful attachment with a preadolescent.

Adult Male with Infant

Our adult males (both feral and isolate reared) who reared infants from one to seven months of age have been separated from their adopted infants in a six-day separation paradigm. Especially in our adult male-male infant dyads, we found an *increasingly* intense attachment over time on the part of the adult male (see Mitchell et al., 1974; Redican et al., 1974).

As we have repeatedly done with other dyads, we examined responses to separation and reunion to evaluate the characteristics of the attachment between the individuals of these pairs (see Plate 5a). One particular question posed in the adult male-infant dyad was whether or not the responses to separation and reunion were similar to those of mother-infant dyads.

Feral-Born Adult Males. Redican found that mother-infant and adult male-infant pairs show similarities and differences in response to separation. Immediately following separation, animals of both types of dyads apparently orient toward the barrier separating them. Both female-reared and male-reared infants are usually more distressed than their adult cagemates. For example, infants give distress vocalizations very frequently and for relatively long durations, whereas adults remain primarily silent; infants are very active, adults much more stationary. The infants are much more active than the males or mothers in trying to reach the other animal through the barrier.

The mothers' emotional responses to separation in the Brandt et al. (1972) study were shown primarily through frequent facial expressions (e.g., retrieval grimaces). Redican's adult males displayed no such expressions. In general, adult males' distress behaviors at separation from an infant were of lesser magnitude than mothers. Individual variability, however, was astounding! One adult male bit himself severely moments after separation. This response was never seen in mothers. Males displayed more tension or conflict than did mothers. It appeared that a mother's response to separation might be characterized as primarily oriented toward the retrieval of her young, whereas adult males were more likely to direct aggression toward the cause of the separation.

All male-infant pairs immediately established contact at reunion, as did most mother-infant pairs. But overall levels of contact in adult male-infant dyads were lower following reunion than before separation. This is in direct contrast to mother-infant pairs. Self-play and interactive play increased among male-infant pairs following reunion, but diminished in mother-infant pairs. It appears that mother-reared infants respond to the stress of separation with increased proximity-oriented responses, whereas male-reared infants initially maintain proximity but then engage in exploratory or arousal-increasing play behaviors afterward. The relatively intense levels of play and other active behaviors engaged in by male-infant dyads during rearing may have contributed to this development, but more detailed analyses are needed.

Isolation-Reared Males. We will now describe a separation project conceived and carried out by Gomber (1975). This project involved separating three adult isolate-reared males from infants they had reared. As in the feral-born adult male-infant dyads, both the isolate-reared adult males and the infants who grew up with them showed signs of distress upon separation, although the infants' reactions were more pronounced and prolonged than were those of the isolate-reared males (the infants were seven-months old at separation).

While overall levels of infant locomotion did not change during separation, there were changes in the infant activities which constituted the locomotion scores. Self-play in the infants, which was moderate during preseparation, dropped out entirely in separation, and reappeared at much higher levels during reunion. Infant locomotor activity on the first day of separation consisted of repeated attempts to get to the isolate-reared adult male. In fact, one female infant sustained facial cuts and scratches from flinging herself headlong at the barrier in attempts to reunite herself with her bizarre adult male "parent." However, infant locomotion scores during the second day of separation were usually accounted for by stereotyped pacing and running, behaviors not seen in the infants during preseparation and reunion. Signs of developing depression were rare.

As in the mother-infant and feral male-infant controls, the isolate-reared adult males were generally less active than were their infants. The isolate adult males varied in their locomotor activity: one increased during separation and reunion, one decreased, and one showed very low levels overall. Each adult male, however, showed higher levels of his most usual stereotyped behavior during the separation stage than during preseparation or reunion. Both the adult isolate males and their infants showed large increases in self-directed behaviors during the separation stage. Mother-infant control pairs also showed an increase in these activities (see Brandt et al., 1972).

In contrast to mothers, but like the feral-born males, adult isolation-reared males did not display facial expressions very frequently in response to separation. Gomber found that the infants responded in much the same way as did mother-reared infants: screeches and coos increased from zero in preseparation to sustained high levels during separation and decreased to a few in the reunion stage.

In general, the infants spent much more time looking at their partners than did the isolate-reared adult males, but the responses of both adults and infants during the three stages varied in the same way. Duration of looking at the partner was fairly constant during preseparation and separation but increased dramatically during reunion. Gomber has suggested that this might reflect increases in attachment at reunion, as a direct effect of having been separated.

There was no ventral-ventral contact in these dyads, in marked contrast to the mother-infant pairs. Unlike the control male-infant dyads, there was no reinstitution of prolonged contact immediately upon reunion. Prolonged contact was eventually reestablished between the adult isolate males and their infants, but the first fifteen to thirty minutes after reunion was spent primarily in energetic self-play by the infants. Whereas contact and grooming showed increased durations during reunion for the mother-infant pairs, the isolate male-infant scores for contact remained at or below preseparation levels, and grooming increased only slightly during reunion. In both groups and in all stages, the infants approached the adults more than the adults approached the infants.

Gomber's analyses show that the responses of isolate male-reared infants to separation and reunion were in some respects similar to those of mother-reared infants, but in many respects were even more like the feral-born male-reared infants. This is particularly striking given the vast differences in early rearing experience and later social competency between the two groups of adult males. Gomber's findings suggest that sex differences in "parenting" potential prevail despite extremely varied early experience in the "parents."

Adult Male with Adult Female

Prior to a study reported by us (Maple et al., 1973a), a generally accepted age for the onset of sexual maturity for the rhesus monkey

(*Macaca mulatta*) had been 4.5 years for males and 3.5 years for females (Napier and Napier, 1967). In our laboratory, however, conception was accomplished at approximately 3.30 years for two laboratory-reared males and 3.11 and 3.13 years for two laboratory-reared females. The onset of sexual maturity in our two males was over one year earlier than that reported by Napier and Napier (1967). In laboratories, there appears to be a humanlike "secular trend" in which animals in captivity reach puberty at earlier and earlier ages (Trollope and Blurton-Jones, 1975). Whatever the usual age of sexual maturity, we have studied the response to separation in both young and fully mature adult heterosexual dyads. We have also published two studies on heterosexual dyads whose ages bordered on sexual maturity (Erwin et al., 1973a; Erwin and Mitchell, 1975). We have already referred to these studies in the sections on adolescent pairings and on adolescent separations, respectively. Plate 5b shows a typical response to heterosexual reunion after two years of separation.

In our initial project on adult heterosexual attachment, separation, and reunion, four pairs of adults were studied (Maple et al., 1973a). All subjects were feral born, but they had lived in a laboratory setting for several years. Each pair was observed during two days each of preseparation, separation, and reunion. The first two days were devoted to obtaining base rates for behavior related to the formation of sexual and emotional attachments. This period was followed by two days in which the animals were separated. Two days of reunion ended the observation period for a given pair.

Few vocalizations were emitted by the adults, but those that were heard were given by females during the separation phase. Lipsmacking occurred more in the males than in the females. Threats were directed toward observers more frequently during preseparation and separation than during reunion, and sexual behavior was seen most frequently during preseparation.

Although the separation effects of two adult monkeys are in many ways different from the separation effects in dyads of other ages, there are similarities. All monkey dyads respond to separation with some form of protest. This protest differs according to age and sex, and many other factors. Interestingly, all four of the feral-reared males bit themselves during the separation phase of this study! Emotional disturbance may be inferred. The females also appeared disturbed, but less so than the males. The animal which is most active in establishing the bond (in our study, the males) may be the one that is most affected by the breaking of the bond. There may also be an effect related to the dominance of the male over the female. We did not examine these two possibilities.

There was one very important factor not controlled in our initial study of adult heterosexual separation–sexual cycle phase. The state of receptivity in the female might have influenced our findings. In an effort to measure the influence of relative receptivity on adult heterosexual separation responses, we did another experiment like the first one but with two separate groups of females: those near ovulation and those menstruating (Maple et al., 1974b).

Unlike the males in our previous experiment, these males did not bite themselves at separation. We have no idea why there was a difference in this behavior. Different males were used, but both groups of males were feral born. Both males and females increased exploration of the cage during separation, but this effect was twice as great in the males and females of the menstruation group. Also during separation, the females cooed while the males did not. Stereotyped and bizarre behaviors were seen in both sexes and in both groups at separation, indicating disturbance.

The males rarely threatened, but females often did. The female threats peaked, however, during the first day of preseparation. They then tapered off and remained very consistent. Females did not threaten males; the threats were directed toward human observers.

During reunion, females' looking at the male partner increased (relative to preseparation). In all cases the reunion level of looking was twice the preseparation level of looking, and in every case but one, the females looked at their males ten times more than the males looked at the females. Follicular-phase females looked at males twice as much as did menstruating females. In both males and females, grooming increased from preseparation to reunion (although sexual behavior did not). Grooming was twice as high in the follicular-phase females as in the menstruating females.

We suspect that the increases in looking and grooming at reunion were signs of heterosexual affection. The fact that sexual behavior was not more frequent at reunion than at preseparation (even in estrous females) convinced us that attachment in adult heterosexual dyads involved more than sexual attraction (also see Agar and Mitchell, 1975). There is potential in the heterosexual adult rhesus dyad for very strong social-emotional bonds.

Adult Male with Adult Male

We have no systematic data on the separation of two adult males. However, Erwin and Maple (1976) described the reunion behavior of two young adult male rhesus monkeys which had become very attached to one another when they were juveniles. The two males spent their entire second year of life alone together. When paired with females during adolescence, both males exhibited normal heterosexual mounting. On several occasions, however, when the two males were reunited with each other, homosexual behavior was displayed. In one instance the two males were reunited after two years of separation. During this reunion and during others, there was apparent recognition and reciprocal mounting, including reciprocal anal penetration. To assess partner preference, the two males were introduced into a cage containing a sexually receptive female (see Erwin, this volume). Both males exhibited aggression toward the female, yet directed sexual and grooming behaviors toward each other (Plate 6). Just as adult rhesus heterosexual attachment involves both asexual and sexual affection, so does adult rhesus like-sex attachment. Agar and Mitchell (1975) have reviewed reports of adult attachment in the wild and Maple (1977) has reviewed studies of unusual sexual behavior in nonhuman primates, including *Macaca mulatta*. In these reviews, and in our own laboratory studies, there is evidence of a potential for fairly intense adult homosexual affection in both the adult male and adult female dyad.

Interspecific Dyads

The four juvenile-aged baboon-macaque pairs studied by Maple (1974a) were also observed in two-day separation paradigms. Separation tests, conducted after six months of interspecies pair living, revealed some differences in the emotional responses of baboons as compared to rhesus monkeys (Maple and Mitchell, 1974). Vocal distress calls, for example, differed in the order of thirty to one in favor of baboons in one study. However, the separation syndrome itself was remarkably similar for baboon and rhesus, and the two members of each interspecific pair were clearly distressed when separated from one another (see Plate 5c). Thus, the distressing effects of separation, it seems, can be generalized across genera within the primate order.

Infant galagos (Sauer, 1967) and lemurs (Jolly, 1967) both emit distress calls when their mothers are not visible. So do squirrel monkeys (Vandenbergh, 1966), vervets (Struhsaker, 1967), langurs (Jay, 1965), and chimpanzees (Goodall, 1967). The effects are widespread in the primate order (cf. Agar and Mitchell, 1973).

DISCUSSION AND CONCLUDING REMARKS

In this chapter we have examined a wide range of social and emotional attachment capabilities in one species of primate, *Macaca mulatta.* At the outset we listed many important research projects of relevance to our own work, including nonprimate and human research contributions as well as nonhuman primate research.

In our second major section, we reviewed extant data on the means of social and emotional communication utilized by this species. In this second section, we described some of the vocalizations, movements, postures, and facial expressions displayed by the rhesus monkey.

We followed our descriptions of rhesus social communication with a detailed report on the nature of attachment in ten different dyads of monkeys, including interspecific dyads of baboons and macaques. These ten dyads differed from one another primarily with regard to the age, sex, and rearing of the individuals involved.

In our final section of this report, we presented data on attachment disruption for the ten dyads discussed above. The effects of age, sex, rearing, and species were reexamined. We also discussed the role of individual differences and the possibilities of generalizability within the primate order.

Our research has provided basic empirical evidence for the combined roles of maturation and social interaction in the development of social and emotional potential. At the least, we have collected developmental data needed for potential application to psychological problems involving the rhesus macaque in captivity.

It is our hope that the information contained here will provide the basis for increased understanding of the nature of social-emotional bonds and their importance for rhesus macaques. We also hope that increased appreciation for attachment processes and potential among captive rhesus monkeys will lead to a more general wisdom and empathy for all animals in captivity. It is clear that the mere provision of companionship (see Maple, Chapter 9, this volume) may not be sufficient to ensure that captive primates develop or maintain psychological health. There can be no doubt that the potential for forming emotional bonds exists across age and sex classes; even rearing history and species differences do not preclude bonding. Nevertheless, some kinds of bonds are more stable than others, and the disruption of some attachments (a common occurrence in breeding colonies and other captive circumstances) entails greater risk than does that of others. Perhaps the research reported here will assist those responsible for making decisions regarding separation or reunion of captive primates.

There is a clear need for further study of attachment potential in applied, as well as research settings. We wish to encourage reports of relevant instances of successful or unsuccessful attempts to pair, separate, or reunite nonhuman primates in captivity. The information will be especially useful if it is quantitative in nature, but purely descriptive reports are also valuable. It is our hope that behavioral research in zoological parks and domestic breeding colonies will be encouraged and funded for its practical applications, as well as its theoretical value.

The ultimate goal of the research reported here was not, however, its application to problems of animal husbandry. We initiated this research because we believe that rhesus monkeys share with humans some fundamental psychological characteristics, and that studies of the kind reported here can contribute by analogy as well as homology to the understanding and/or elimination of human ills and misfortunes, including crime, violence, drug addiction, sexual deviations, alcoholism, and suicide. Emotional protest, aberrant behavior, alienation, and psychological depression are all correlates or effects of the processes of psychological loss of human objects of attachment. The separation from or rejection by objects of love, whether real or imagined, can be personally devastating, as each of us knows too well. It is not at all surprising that in many violent crimes, the victim and perpetrator are emotionally attached to one another. The distress of separation releases deep and serious emotional arousal which may underlie many of the psychological and sociological problems associated with sexual and agonistic behavior.

REFERENCES

Agar, M. E. and Mitchell, G. A bibliography on deprivation and separation with special emphasis on primates. *Catalog of Selected Documents in Psychology 3* (1973) 20 pp.

Agar, M. E. and Mitchell, G. Behavior of free-ranging adult rhesus macaques. In G. Bourne (Ed.) *The Rhesus Monkey*, Vol. I. *Anatomy and Physiology.* New York: Academic Press, 1975, pp. 323–342.

Altmann, M. Naturalistic studies of maternal care in moose and elk. In H. L. Rheingold (Ed.) *Maternal Behavior in Mammals.* New York: Wiley, 1963, pp. 233–253.

Anderson, C. O. and Mason, W. A. Early experiences and complexity of social organizations in groups of young rhesus monkeys (*Macaca mulatta*). *J. Comp. Physiol. Psychol.* **87**: 681–690 (1974).

Bateson, P. P. G. The development of social attachments in birds and man. *Adv. Sci.* **25**: 279–288 (1969).

Bayley, N. The development of motor abilities during the first three years. *Monogr. Soc. Res. Child Dev.* **1**: 1–26 (1935).

Baysinger, C., Brandt, E. M., and Mitchell, G. Development of infant rhesus monkeys (*Macaca mulatta*) in their isolation environments. *Primates* **13**: 257–270 (1972).

Berkson, G. Abnormal stereotyped motor acts. In J. Zubin and H. F. Hunt (Eds.) *Comparative Psychopathology.* New York: Grune & Stratton, 1967, pp. 76–94.

Bo, W. J. Parturition. In E. S. E. Hafez (Ed.) *Comparative Reproduction of Nonhuman Primates.* Springfield, Ill.: Charles C Thomas, 1971, pp. 302–314.

Bowlby, J. The nature of the child's tie to the mother. *Int. J. Psychoanal.* **41**: 89–113 (1958).

Bowlby, J. Separation anxiety. *Int. J. Psychoanal.* **41**: 89–113 (1960).

Bowlby, J. *Attachment and Loss*, Vol. 1. *Attachment.* New York: Basic Books, 1969.

Bowlby, J. *Attachment and Loss*, Vol 2. *Separation.* New York: Basic Books, 1973.

Brandt, E. M. and Mitchell, G. Parturition in primates: Behavior related to birth. In L. A. Rosenblum (Ed.) *Primate Behavior: Developments in Field and Laboratory Research*, Vol. 2. New York: Academic Press, 1971, pp. 177–223.

Brandt, E. M. and Mitchell, G. Pairing preadolescents with infants (*Macaca mulatta*). *Dev. Psychol.* **8**: 222–228 (1973a).

Brandt, E. M. and Mitchell, G. Labor and delivery behavior in rhesus monkeys (*Macaca mulatta*). *Am. J. Phys. Anthropol.* **38**: 519–522 (1973b).

Brandt, E. M., Baysinger, C., and Mitchell, G. Separation from rearing environment in mother-reared and isolation-reared rhesus monkeys (*Macaca mulatta*). *Int. J. Psychobiol.* 2: 193–204 (1972).

Brandt, E. M., Stevens, C. W., and Mitchell, G. Visual social communication in adult male isolate-reared monkeys (*Macaca mulatta*). *Primates* **12**: 105–112 (1971).

Bronson, G. W. The development of fear in man and other animals. *Child Dev.* **39**: 409–431 (1968).

Buckley, W. (Ed.) *Modern Systems Research for the Behavioral Scientists.* Chicago: Aldine, 1968.

Cairns, R. B. Attachment behavior of mammals. *Psychol. Rev.* **73**: 409–426 (1966).

Cairns, R. B. Attachment and dependency: A psychobiological and social learning synthesis. In J. L. Gewirtz (Ed.) *Attachment and Dependency.* New York: Winston & Sons, 1972, pp. 29–80.

Carpenter, C. R. Sexual behavior of free ranging rhesus monkeys (*Macaca mulatta*). *J. Comp. Psychol.* **33**: 113–162 (1942).

Casler, L. R. Maternal deprivation: A critical review of the literature. *Monogr. Soc. Res. Child Dev.* **26**: no. 80 (1961).

Coulson, J. C. The influence of the pair bond and age on the breeding biology of the kittiwake gull (*Rissa tridactyla*). *J. Anim. Ecol.* **35**: 269–279 (1966).

Cross, H. A. and Harlow, H. F. Prolonged and progressive effects of partial isolation on the behavior of macaque monkeys. *J. Exp. Res. Personality* **1**: 39–49 (1965).

Darwin, C. *The Expressions of the Emotions in Man and Animals.* London: J. Murray, 1872.

Dember, W. N. The new look in motivation. *Am. Sci.* **53**: 409–427 (1965).

Denenberg, V. H. Stimulation in infancy, emotional reactivity, and exploratory behavior. In D. C. Glass (Ed.) *Neurophysiology and Emotion.* New York: Russell Sage Foundation, 1967, pp. 161–189.

Dennis, W. Causes of retardation among institutionalized children: Iran. *J. Genet. Psychol.* **96**: 47–59 (1960).

Engle, G. L. Anxiety and depression-withdrawal: The primary affects of unpleasure. *Int. J. Psychoanal.* **43**: 89–97 (1962).

Erwin, J. The development and persistence of rhesus macaque (*Macaca mulatta*) peer attachments. Ph.D. dissertation, University of California, Davis, 1974a.

Erwin, J. Responses of rhesus monkeys to the separation vocalizations of a conspecific infant. *Percept. Mot. Skills* **39**: 179–185 (1974b).

Erwin, J. Rhesus monkey vocal sounds. In G. Bourne (Ed.) *The Rhesus Monkey*, Vol. 1. *Anatomy and Physiology.* New York: Academic Press, 1975, pp. 365–380.

Erwin, J., and Flett, M. Responses of rhesus monkeys to reunion after long-term separation: Cross-sexed pairings. *Psychol. Rep.* **35**: 171–174 (1974).

Erwin, J. and Maple, T. Ambisexual behavior in a pair of laboratory-reared male rhesus monkeys. *Arch. Sex. Behav.* **5**: 9–14 (1976).

Erwin, J. and Mitchell, G. Analysis of rhesus monkey vocalizations: Maturation-related changes in clear call frequency. *Am. J. Phys. Anthropol.* **38**: 463–468 (1973).

Erwin, J. and Mitchell, G. Initial heterosexual behavior of adolescent rhesus monkeys (*Macaca mulatta*). *Arch. Sex. Behav.* **4**: 97–104 (1975).

Erwin, J., Brandt, E. M., and Mitchell, G. Attachment formation and separation in heterosexually naive pre-adult rhesus monkeys. *Dev. Psychobiol.* **6**: 531–538 (1973a).

Erwin, J., Maple, T., and Welles, J. Responses of rhesus monkeys to reunion: Evidence for exclusive and persistent bonds between peers. *Contemporary Primatology*. Basel: Karger, 1975, pp. 254–262.

Erwin, J., Mitchell, G., and Maple, T. Abnormal behavior in non-isolate reared rhesus monkeys. *Psychol. Rep.* **33**: 515–523 (1973b).

Erwin, J., Mobaldi, J., and Mitchell, G. Separation of juvenile rhesus monkeys of the same sex. *J. Abnorm. Psychol.* **78**: 134–139 (1971).

Erwin, J., Maple, T., Mitchell, G., and Willott, J. A follow-up study of isolation- and mother-reared rhesus monkeys which were paired with preadolescent conspecifics in late infancy: Cross-sexed pairings. *Dev. Psychol.* **10**: 423–428 (1974a).

Erwin, J., Maple, T., Willott, J. F., and Mitchell, G. Persistent peer attachments in rhesus monkeys: Responses to reunion after two years of separation. *Psychol. Rep.* **34**: 1179–1183 (1974b).

Fantz, R. L. Pattern vision in newborn infants. *Science* **140**: 296–297 (1963).

Fittinghoff, N., Lindburg, D. G., Gomber, J., and Mitchell, G. Consistency and variability in the behavior of mature, isolate-reared male rhesus macaques. *Primates* **15**: 111–139 (1974).

Fragaszy, D. M. and Mitchell, G. Infant socialization in macaques. *J. Hum. Evol.* **3**: 430–439 (1974).

Freedman, D. G. The infants' fear of strangers and the flight response. *J. Child Psychol. Psychiatry* **2**: 242–248 (1961).

Freud, S. *New Introductory Lectures on Psychoanalysis*, transl. by W. J. H. Sprott. New York: Norton, 1933.

Gallup, G. G. and McClure, M. K. Preference for mirror image stimulation in differentially reared rhesus monkeys. *J. Comp. Physiol. Psychol.* **75**: 403–407 (1971).

Gareffa, L. F. Group membership and group size as determinants of aggressive behavior between pairs of male bobwhite quail (*Colinus virginianus*). *Commun. Behav. Biol.* **3**: 69–72 (1969).

Gewirtz, J. L. Attachment and dependence: Some strategies and tactics in selection and use of indices for those concepts. In *Communication and Affect*. New York: Academic Press, 1972, pp. 19–49.

Goldfarb, W. Psychological privation in infancy and subsequent adjustment. *Am. J. Orthopsychiatry* **15**: 247–255 (1945).

Gomber, J. Caging adult male isolation-reared rhesus monkeys (*Macaca mulatta*) with infant conspecifics. Ph.D. dissertation, University of California, Davis, 1975.

Gomber, J. and Mitchell, G. Preliminary report on adult male isolation-reared rhesus monkeys caged with infants. *Dev. Psychol.* **9**: 419 (1974).

Goodall, J. van Lawick. Mother-offspring relationships in free-ranging chimpanzees. In D. Morris (Ed.) *Primate Ethology.* Chicago: Aldine, 1967, pp. 287–346.

Griffin, G. A. The effects of multiple mothering on the infant-mother and infant-infant affectional systems. Unpublished Ph.D. dissertation, University of Wisconsin, Madison, 1966.

Grimm, R. J. Catalog of sounds of the pigtailed macaque (*Macaca nemestrina*). *J. Zool. London* **152**: 361–373 (1967).

Hansen, E. W. The development of maternal and infant behavior in the rhesus monkey. *Behaviour* **27**: 107–149 (1966).

Harlow, H. F., Harlow, M. K., and Hansen, E. W. The maternal affectional system of rhesus monkeys. In H. L. Rheingold (Ed.) *Maternal Behavior in Mammals.* New York: Wiley, 1963, pp. 254–281.

Harms, J. W. Schwangerschaft und geburt. In H. Hofer, A. H. Schultz, and D. Starck (Eds.) *Primatologia: A Handbook of Primatology*, Vol. 1. New York: S. Karger, 1956, pp. 661–772.

Harper, L. V. Ontogenetic and phylogenetic functions of the parent-offspring relationship in mammals. *Advances in the Study of Behavior* **3**: 75–117 (1970).

Hartman, C. G. The period of gestation in the monkey, *Macacus rhesus*, first description of parturition in monkeys, size and behavior of young. *J. Mammals* **9**: 181 (1928).

Hausfater, G. Intergroup behavior of free-ranging rhesus monkeys (*Macaca mulatta*). *Folia Primatol.* **18**: 78–107 (1972).

Heinicke, C. and Westheimer, I. J. *Brief Separations.* New York: International University Press, 1965.

Hinde, R. A. and Rowell, T. E. Communication by posture and facial expression in the rhesus monkey. *Proc. Zool. Soc. London* **138**: 1–21 (1962).

Hinde, R. A., Spencer-Booth, Y., and Bruce, M. Effects of six day maternal deprivation on rhesus monkey infants. *Nature* **10**: 1021–1023 (1966).

Itani, J. Vocal communication of the wild Japanese monkey. *Primates* **4**: 11–66 (1963).

Jacobson, H. N. and Windle, W. F. Observations on mating, gestation, birth, and postnatal development of *Macaca mulatta. Biol. Neonat.* **2**: 105–120 (1960).

Jay, P. The common langur of North India. In I. DeVore (Ed.) *Primate Behavior.* New York: Holt, Rhinehart and Winston, 1965, pp. 197–249.

Jensen, G. D. and Tolman, C. W. Mother-infant relationship in the monkey, *Macaca nemestrina*: The effect of brief separation and mother-infant specificity. *J. Comp. Physiol. Psychol.* **55**: 131–136 (1962).

Jolly, A. G. *Lemur Behavior: A Madagascar Field Study.* Chicago: University of Chicago Press, 1967.

Kahmann, H. and Frisch, O. V. The relations between mother animal and baby in small mammals (transl. from *Experimentia* **8**: 221–223 1952).

Kaufman, I. C. and Rosenblum, L. A. Depression in infant monkeys separated from their mothers. *Science* **155**: 1030–1031 (1967a).

Kaufman, I. C. and Rosenblum, L. A. The reaction to separation in infant monkeys: Anaclitic depression and conservational withdrawal. *Psychosom. Med.* **29**: 648–676 (1967b).

King, D. L. A review and interpretation of some aspects of the infant-mother relationship in mammals and birds. *Psychol. Bull.* **65**: 143–155 (1966).

Koford, C. B. Population changes in the rhesus monkeys of Cayo Santiago (1960–1964). *Tulane Studies in Zool.* **13**: 1–17 (1966).

Koford, C. B., Farber, P. A., and Windle, W. F. Twins and teratisms in the rhesus monkey. *Folia Primatol.* **4**: 221–222 (1966).

Kummer, H. *Primate Societies: Group Techniques of Ecological Adaptation.* Chicago: Aldine, 1971.

Kuo, Z. Y. Studies on the basic factors in animal fighting: V. Interspecies coexistence in fish. *J. Genet. Psychol.* **97**: 181–194 (1960a).

Kuo, Z. Y. Studies on the basic factors in animal fighting: VI. Interspecies coexistence in birds. *J. Genet. Psychol.* **97**: 195–209 (1960b).

Kuo, Z. Y. Studies on the basic factors in animal fighting: VII. Interspecies coexistence in mammals. *J. Genet. Psychol.* **97**: 211–225 (1960c).

Kuo, Z. Y. *The Dynamics of Behavior Development: An Epigenetic View.* New York: Random House, 1967.

Lack, D. Pair formation in birds. *Condor* **42**: 269–286 (1940).

LeResch, R. E. and Sladen, W. J. L. Establishment of pair and breeding site bonds by young known-age adelie penguins (*Pygoscelis adeliae*). *Anim. Behav.* **18**: 517–526 (1970).

Levine, S. Stress and behavior. In R. F. Thompson (Ed.) *Physiological Psychology Readings from Scientific American.* San Francisco: Freeman, 1971, pp. 195–198.

Lindburg, D. G. The rhesus monkey in North India: An ecological and behavioral study. In L. A. Rosenblum (Ed.) *Primate Behavior: Developments in Field and Laboratory Research*, Vol. 2. New York: Academic Press, 1971, pp. 1–106.

Lindburg, D. G. and Hazell, L. D. Licking of the neonate and duration of labor in great apes and man. *Am. Anthropol.* **74**: 318–325 (1972).

Lorenz, K. Z. The companion in the birds' world. *Auk* **54**: 245–273 (1937).

Maple, T. Behavioral comparisons of infant baboons housed with infant macaques in a laboratory environment. In *Abstracts of Symposia and Contributed Papers of the 54th Meeting of the Western Society of Naturalists*, San Diego, Dec. 1973, p. 17.

Maple, T. Raising baboons with macaques. *The Simian* May: 3–4 (1974a).

Maple, T. Basic studies of interspecies attachment behavior. Ph.D. dissertation, University of California, Davis, 1974b.

Maple, T. On the need for investigations of interspecies social behavior within the order of primates. *J. Behav. Sci.* **2**: 63–66 (1974c).

Maple, T. Unusual sexual behavior of nonhuman primates. In J. Money and H. Musaph (Eds.) *Handbook of Sexology*. Amsterdam: Elsevier, 1977, pp. 1167–1186.

Maple, T. and Lawson, R. Interspecies vocal communication between a baboon and a macaque. *Primates* **16**: 99–101 (1975).

Maple, T. and Mitchell, G. Behavioral responses to separation in interspecific pairs of baboons and macaques. *Am. Zool.* **14**: 224 (1974).

Maple, T., Brandt, E. M., and Mitchell, G. Separation of preadolescents from infants (*Macaca mulatta*). *Primates* **16**: 141–153 (1975).

Maple, T., Erwin, J., and Mitchell, G. Age of sexual maturity in laboratory born pairs of rhesus monkeys (*Macaca mulatta*). *Primates* **14**: 427–428 (1973a).

Maple, T., Erwin, J., and Mitchell, G. Sexually aroused self-aggression in a socialized adult male monkey. *Arch. Sex. Behav.* **3**: 471–475 (1974a).

Maple, T., Erwin, J., and Mitchell, G. Separation of adult heterosexual pairs of rhesus monkeys: The effects of female cyclicity. *J. Behav. Sci.* **2**: 81–86 (1974b).

Maple, T., Risse, G., and Mitchell, G. Separation of adult male from adult female rhesus monkeys (*Macaca mulatta*). *J. Behav. Sci.* **1**: 327–336 (1973b).

Mason, W. A. The effects of environmental restriction on the social development of rhesus monkeys. In. C. H. Southwick (Ed.) *Primate Social Behavior.* Princeton, N.J.: Van Nostrand, 1963a, pp. 161–173.

Mason, W. A. Social development of rhesus monkeys with restricted social experience. *Percept. Mot. Skills* **16**: 263–270 (1963b).

Mason, W. A. Early social deprivation in the nonhuman primates: Implications for human behavior. In D. C. Glass (Ed.) *Environmental Influences.* New York: Rockefeller University and Russell Sage, 1968, pp. 70–100.

Mason, W. A. Motivational factors in psychosocial development. In W. H. Arnold and M. M. Page (Eds.) *Nebraska Symposium on Motivation.* Lincoln: University of Nebraska Press, 1971.

Mason, W. A. and Kenney, M. D. Redirection of filial attachments in rhesus monkeys: Dogs as mother surrogates. *Science* **183**: 1209–1211 (1974).

McKinney, W. T., Suomi, S. J., and Harlow, H. F. Repetitive peer separations of juvenile age rhesus monkeys. *Arch. Gen. Psychiatry* **27**: 200–203 (1972).

Meier, G. W. Behavior of infant monkeys: Differences attributable to mode of birth. *Science* **143**: 968–970 (1964).

Meier, G. W. Maternal behavior of feral and laboratory-reared monkeys following the surgical delivery of their infants. *Nature* (*London*) **206**: 402–493 (1965).

Michael, R. P. and Zumpe, D. Patterns of reproductive behavior. In E. S. E. Hafez (Ed.) *Comparative Reproduction of Nonhuman Primates.* Springfield, Ill.: Charles C Thomas, 1971, pp. 205–242.

Miller, R. E., Caul, W. F., and Mirsky, I. A. Communication of affects between feral and socially isolated monkeys. *J. Personal. Soc. Psychol.* **7**: 231–240 (1967).

Mitchell, G. Persistent behavior pathology in rhesus monkeys following early social isolation. *Folia Primatol.* **8**: 132–147 (1968a).

Mitchell, G. Attachment differences in male and female infant monkeys. *Child Dev.* **39**: 611–120 (1968b).

Mitchell, G. Abnormal behavior in primates. In L. A. Rosenblum (Ed.) *Primate Behavior: Developments in Field and Laboratory Research*, Vol. 1. New York: Academic Press, 1970, pp. 195–249.

Mitchell, G. Parental and infant behavior. In E. S. E. Hafez (Ed.) *Comparative Reproduction of Laboratory Primates.* Springfield, Ill.: Charles C Thomas, 1971, pp. 382–402.

Mitchell, G. Looking behavior in the rhesus monkey. *J. Phenomenolog. Psychol.* **3**: 53–67 (1972).

Mitchell, G. Parental behavior in nonhuman primates. In J. Money and H. Musaph (Eds.) *Handbook of Sexology.* Amsterdam: Elsevier, 1977, pp. 749–759.

Mitchell, G. and Brandt, E. M. Behavioral differences related to experience of mother and sex of infant in rhesus monkey. *Dev. Psychol.* **3**: 149 (1970).

Mitchell, G. and Brandt, E. M. Behavior of the rhesus monkey at birth. In G. Bourne (Ed.) *The Rhesus Monkey*, Vol. 2. New York: Academic Press, 1975, pp. 231–244.

Mitchell, G. and Redican, W. K. Communication in normal and abnormal rhesus monkeys. *Proceedings of the XXth International Congress of Psychology*, Tokyo, Japan, Science Council of Japan, 1972, pp. 171–172.

Mitchell, G. and Schroers, L. Birth order and parental experience in monkeys and man. In H. W. Reese (Ed.) *Advances in Child Development and Behavior*, Vol. 8. New York: Academic Press, 1973, pp. 159–184.

Mitchell, G. and Stevens, C. W. Primiparous and multiparous monkey mothers in a mildly stressful social situation: I. First three months. *Dev. Psychol.* **1**: 280–286 (1969).

Mitchell, G., Redican, W. K., and Gomber, J. Lesson from a primate: Males can raise babies. *Psychol. Today* **7**: 63–68 (1974).

Mitchell, G., Harlow, H. F., Griffin, G. A., and Mφller, G. W. Repeated maternal separation in the monkey. *Psychonom. Sci.* **8**: 197–198 (1967).

Mitchell, G., Raymond, E., Ruppenthal, G., and Harlow, H. Long-term effects of total social isolation upon behavior of rhesus monkeys. *Psychol. Rep.* **18**: 567–580 (1966a).

Mitchell, G., Ruppenthal, G. C., Raymond, E. J., and Harlow, H. F. Long term effects of multiparous and primiparous monkey mother rearing. *Child Dev.* **37**: 781–791 (1966b).

Moltz, H. Imprinting: An epigenetic approach. *Psychol. Rev.* **70**: 123–138 (1963).

Morris, R. L. and Erickson, C. J. Pair bond maintenance in the ring dove (*Streptopelia resoria*). *Anim. Behav.* **19**: 398–406 (1971).

Napier, J. and Napier, P. *A Handbook of Living Primates.* New York: Wiley, 1967.

Neville, M. K. A study of the free-ranging behavior of rhesus monkeys. Ph.D. dissertation, Department of Anthropology, Harvard University, Cambridge, Mass., 1966.

Orians, G. H. On the evolution of mating systems in birds and mammals. *Am. Nat.* **103**: 589–603 (1969).

Redican, W. K. Facial expressions in nonhuman primates. In L. A. Rosenblum (Ed.) *Primate Behavior: Developments in Field and Laboratory Research*, Vol. 4. New York: Academic Press, 1975, pp. 103–194.

Redican, W. K. and Mitchell, G. The social behavior of adult male-infant pairs of rhesus monkeys in a laboratory environment. *Am. J. Phys. Anthropol.* **38**: 523–526 (1973).

Redican, W. K. and Mitchell, G. Play between adult male and infant rhesus monkeys. *Am. Zool.* **14**: 295–302 (1974).

Redican, W. K., Gomber, J., and Mitchell, G. Adult male parental behavior in feral- and isolation-reared rhesus monkeys (*Macaca mulatta*). In J. H. Cullen (Ed.) *Experimental Behaviour: A Basis for the Study of Mental Disturbance.* Dublin: Irish University Press, 1974, pp. 131–146.

Redican, W. K., Kellicutt, M. H., and Mitchell, G. Preference for facial expressions in rhesus monkeys. *Dev. Psychol.* **5**: 539 (1971).

Rheingold, H. L. (Ed.) *Maternal Behavior in Mammals.* New York: Wiley, 1963.

Rheingold, H. L. and Bayley, N. The later effects of an experimental modification of mothering. *Child Dev.* **30**: 362–372 (1959).

Rosenblum, L. A. and Kaufman, I. C. Variations in infant development and response to maternal loss in monkeys. *Am. J. Orthopsychiatry* **28**: 418–426 (1968).

Rowell, T. E. Agonistic noises of the rhesus monkey (*Macaca mulatta*). *Symp. Zool. Soc. London* **8**: 91–96 (1962).

Rowell, T. E. The social development of some rhesus monkeys. In B. M. Foss (Ed.) *Determinants of Infant Behaviour*, Vol. 2. New York: Wiley, 1963, pp. 35–49.

Rowell, T. E. *Social Behaviour of Monkeys.* Kingsport, Tenn.: Kingsport Press, Inc., 1972.

Rowell, T. E. and Hinde, R. A. Vocal communication by the rhesus monkey (*Macaca mulatta*). *Proc. Zool. Soc. London* **138**: 279–294 (1962).

Rowell, T. E., Hinde, R. A., and Spencer-Booth, Y. "Aunt"-infant interaction in captive rhesus monkeys. *Anim. Behav.* **12**: 219–226 (1964).

Rowland, G. L. The effects of total social isolation upon learning and social behavior in rhesus monkeys. Ph.D. dissertation, University of Wisconsin, Madison, 1964.

Sackett, G. P. Effects of rearing conditions upon the behavior of rhesus monkeys (*Macaca mulatta*). *Child Dev.* **36**: 855–868 (1965).

Salzen, E. A. Imprinting and environmental learning. In L. R. Aronson, D. S.

Lehrman, J. S. Rosenblatt, and E. Tobach (Eds.) *Development and Evolution of Behavior*. New York: Freeman, 1970, pp. 158–180.

Sauer, E. G. Mother-infant relationship in Galagos and the oral child-transport among primates. *Folia Primatol.* **7**: 127–149 (1967).

Schaffer, H. R. Some issues for research in the study of attachment behavior. In B. M. Foss (Ed.) *Determinants of Infant Behavior*, Vol. 2. New York: Wiley, 1963, pp. 179–198.

Schlottman, R. S. and Seay, B. M. Mother-infant separation in *Macaca irus*. Paper presented at the Annual Meeting of the Southeastern Psychological Association, 1968, Roanoke, Va.

Schneirla, T. C. An evolutionary and developmental theory of biphasic processes underlying approach and withdrawal. In M. R. Jones (Ed.) *Nebraska Symposium on Motivation.* Lincoln: University of Nebraska Press, 1959, pp. 1–42.

Scollay, P. Mother-infant separation in rhesus monkeys (*Macaca mulatta*). Ph.D. dissertation, University of California, Davis. 1970.

Scott, J. P. Critical periods in behavioral development. *Science* **138**: 949–958 (1962).

Seay, B. and Harlow, H. F. Maternal separation in the rhesus monkey. *J. Nerv. Ment. Disorders* **140**: 434–441 (1965).

Seay, B., Alexander, B. K., and Harlow, H. F. The maternal behavior of socially deprived rhesus monkeys. *J. Abnorm. Soc. Psychol.* **69**: 345–354 (1964).

Seay, B. M., Hansen, E. W., and Harlow, H. F. Mother-infant separation in monkeys. *J. Child Psychol. Psychiatry* **3**: 123–132 (1962).

Selander, R. K. On mating systems and sexual selection. *Am. Nat.* **99**: 129–141 (1965).

Skeels, H. M. Adult status of children with contrasting early life experiences. *Monogr. Soc. Res. Child Dev.* **31**: no. 105 (1966).

Sluckin, W. *Imprinting and Early Learning.* Chicago: Aldine, 1965.

Spalding, D. A. Instinct, with original observations on young animals. *Macmillan's Magazine* **27**: 282–293 (1873).

Spencer-Booth, Y. and Hinde, R. A. The effects of separating rhesus monkey infants from their mothers for six days. *J. Child Psychol. Psychiatry* **7**: 179–198 (1966).

Spitz, R. A. Motherless infants. *Child Dev.* **20**: 145–155 (1949).

Spitz, R. A. Anxiety in infancy. *Int. J. Psychoanal.* **41**: 89–113 (1950).

Stettner, L. J., Gareffa, L. F., and Missakian, E. Monogamous behavior in the bob-white quail (*Colinus virginianus*). *Psychol. Rev.* **78**: 137–162 (1971).

Stevens, C. W. and Mitchell, G. Birth order effects, sex differences and sex preferences in the peer-directed behavior of rhesus infants. *Int. J. Psychobiol.* **2**: 117–128 (1972).

Struhsaker, T. Auditory communication among vervet monkeys. In S. A. Altmann (Ed.) *Social Communication Among Primates.* Chicago: University of Chicago Press, 1967, pp. 281–324.

Sutton, D., Larson, C., Taylor, E. M., and Lindeman, R. Vocalization in rhesus monkeys: Conditionability. *Brain Res.* **52**: 225–231 (1973).

Tinkelpaugh, O. L. and Hartman, C. G. Behavioral aspects of parturition in the monkey. *J. Comp. Psychol.* **11**: 63–98 (1930).

Trollope, J. and Blurton-Jones, N. G. Aspects of reproduction and reproductive behavior in *Macaca arctoides. Primates* **16**: 191–205 (1975).

Vandenbergh, J. C. Behavioral observations of an infant squirrel monkey. *Psychol. Rep.* **18**: 683–688 (1966).

Werner, H. The concept of development from a comparative and organismic point of view. In D. B. Harris (Ed.) *The Concept of Development.* Minneapolis: University of Minnesota Press, 1957, pp. 125–148.

Windle, W. F. Asphixia at birth, a major factor in mental retardation: Suggestions for prevention based on experiments in monkeys. In J. Zubin and G. Jervis (Eds.) *Psychopathology of Mental Development.* New York: Grune & Stratton, 1967, pp. 140–147.

Wolff, P. H. Observations on newborn infants. *Psychosom. Med.* **21**: 110–118 (1959).

Yarrow, L. J. Attachment and dependency: A developmental perspective. In J. L. Gewirtz (Ed.) *Attachment and Dependency.* New York: Winston & Sons, 1972, pp. 81–95.

Yerkes, R. M. and Elder, J. H. Concerning reproduction in the chimpanzee. *Yale J. Biol. Med.* **10**: 41–48 (1936).

Zuckerman, S. *The Social Life of Monkeys and Apes.* London: Routledge, 1932.

4
Behavior of Primates Present During Parturition*

Nancy Caine and G. Mitchell

Department of Psychology
University of California
Davis, Calif.

INTRODUCTION

The study of parturition in nonhuman primates has largely been confined to descriptions of the mother's behavior and/or physiology during delivery. Studies of group response to birth, especially in natural settings, have been conspicuously absent. One reason for the dearth of information in this area is the inherent difficulty in observing the birth process. Births most often occur at night, precluding observation unless artificial light is available. Such lighting may, in turn, influence the birth process and decrease the validity of the report. Furthermore, births often occur with little or no warning, for it is difficult to estimate the due date of a specific pregnant female unless time of conception is known with some degree of accuracy. Despite these difficulties and limitations, data regarding group responses to birth in several species of nonhuman primates are available. In most cases, however, no more than brief mention is made of the birth-related behavior of conspecifics. Indeed, many of the data on captive birth have been collected from isolated animals.

Regardless of the nature of the data, however, there is a need to compile and organize the information available on group responses to birth in captive and free-ranging nonhuman primates. Such data are of particular importance in light of the current interest in kin selection and parental investment.

The purpose of this chapter, then, is to review and present the information available on the topic of group response to birth in nonhu-

*The authors thank Olivia Scheffler for translations from the German.

man primates. A taxonomic framework is provided to facilitate analysis. Finally, suggestions are made regarding the relevance of the data to primate social systems and to the principles of kin selection and parental investment.

REVIEW OF PRIMATE RESPONSES TO PARTURITION

Prosimians

Tupaiidae. Sorenson and Conaway (1966), in discussing the social bonds between female tree shrews (*Tupaia longipes*), reported that an adult female protected another female while the latter was in labor. The protective female served the purpose of keeping an adult male from entering the nest box in which the laboring female lay. Unfortunately, however, cannibalism or parental desertion is very often the fate of captive newborn tree shrews (Sorenson, 1970).

Lemuridae. One report of parturition in the lemur is that of Basilewsky (1965). The author mentions that all members of captive lemur groups take great interest in parturition, but offers us no further elaboration.

Lorisidae. Doyle et al. (1967) observed several births in the lesser bushbaby, *Galago senegalensis moholi.* In one case the adult male showed increased sexual interest in the female just prior to parturition. Typically, however, the adult males maintained a distance between themselves and the laboring females. Female cagemates showed the most consistent interest in parturition and in the newborn. In one instance an adult female groomed the genital area of a female who was about to deliver. Similarly, a female was once observed to eat a portion of the placenta. In another birth situation the female was not subjected to any interference from her cagemates.

New World Monkeys

Callithricidae. Lucas et al. (1937) were never able to observe a birth in their captive marmosets (*Callithrix jacchus*). However, the authors cited a personal communication which reported paternal assistance in delivery. Langford (1963) likewise reported male assistance in delivery in a group of captive marmosets (*C. jacchus*) housed in a large outdoor enclosure. Males were also observed to clean the neonates before handing the latter to the mothers for feeding.

Rothe (1974) observed the parturition of a marmoset (*C. jacchus*) in the presence of other group members. The female withdrew herself from the group during labor and there was no reported interest on the part of the conspecifics. When the lights were turned on in the previously darkened cage, however, all group members, including an adult male and juvenile male, showed great interest in the neonate. The adult male attempted to copulate with the female, and juveniles of both sexes ate the placenta which had been forcefully seized from the female. The author predicted, however, that placentophagia would have been restricted to the mother had the group been undisturbed by lighting. Rothe reported that the adult males' responses to parturition were extremely variable, ranging from complete indifference to intense concern.

Stevenson (1976) investigated the possibility that marmosets (*C. jacchus*) take characteristic postpartum roles. The birth process was observed seven times in five family groups of captive animals. In some cases the group members shared in placentophagia, although adult males were never seen to do so. In no case did the female isolate herself during parturition, but she would not allow handling of the neonate until one hour after birth, at which time the adult male was allowed such freedom. Likewise, the adult male was the closest of the animals to the female immediately after parturition. All animals were reported to take an active interest in the birth process, regardless of lighting in the cage.

Blakely and Curtis (unpublished manuscript) observed the birth of a captive golden lion marmoset (*Leontopithecus rosalia*) in the presence of the adult male. The male's attention to the female increased with progression of labor; indeed, the male was reported to be in a more highly agitated state than the female. The male repeatedly peered beneath the female and touched the crowning head. He attempted to take the baby but was rebuffed by the mother until she had bitten off the umbilical cord. At this time, the father took the baby and held it.

Cebidae

Howler monkey. Carpenter (1934) observed immediate postpartum behavior on three separate occasions in the wild howler monkey (*Alouatta palliata*). In two of these instances a circle of at least two adult females and/or juveniles surrounded the new mother. These animals were judged to be specifically interested in the infant, as they attempted on several occasions to touch it. The mother avoided these approaches by turning away. Carpenter (1965) reaffirmed these observations in a later study on wild howler monkeys. Neville (1972) was unable to observe parturition in the wild, but did report an interest in a newborn on the part of group females. Males apparently showed no such interest.

Woolly monkey. Williams (1967) reported on the birth of a captive Humboldt's woolly monkey (*Lagothrix lagothrica*) housed with eleven conspecifics. All of the group members gathered around and watched the laboring female intently; the adult male watched with particular concern. No physical interference in the process was noted, however. The male maintained his protective attitude toward the mother and infant immediately after birth.

Saki monkey. Hanif (1967) reported briefly on the birth of a saki monkey (*Pithecia pithecia*) in captivity. The author noted that the adult male attempted to steal the newborn from its mother several times.

Squirrel monkey. Bowden et al. (1967) observed the birth of a squirrel monkey (*Saimiri sciureus*) in a captive group of five conspecifics. As soon as the infant's head crowned, both males retreated to the opposite corner of the cage and remained there during the entire delivery and early postpartum period. However, the young female stayed in the immediate vicinity of the mother throughout delivery. Following parturition the same juvenile female was removed due to her incessant attempts to grab, smell, and tug the infant. In a second observed birth, the same type of interference was exhibited by a juvenile male. The adult male in the cage was seen to retreat from the parturition process. Hopf (1967), however, reported that group interest in the birth of squirrel monkeys occurs only under lighted conditions.

Old World Monkeys

Cercopithecidae

Baboons. Dunbar and Dunbar (1974) witnessed a birth in a free-ranging group of gelada baboons (*Theropithecus gelada*). The female in this case did not isolate herself from the group during parturition, but the circle of animals with which she was sitting fled at the moment of delivery. An infant was the first to approach the mother and neonate; shortly thereafter an adult female and a juvenile female also approached. Only the juvenile was allowed to groom the mother. In the $3\frac{1}{2}$-hour period following parturition a total of fifteen juveniles (mostly female) approached the mother. None was allowed to touch the infant. However, in another postpartum incident a juvenile female was seen to touch a newborn.

Kummer (1968), in his comprehensive field study of hamadryas baboons (*Papio hamadryas*) observed one birth, in spite of the difficulties involved in observing parturition in the field. During the birth the mother sat approximately two meters from the rest of the group. Several adult females approached and sat next to the female and infant, watching intently and following whenever the duo moved.

Bolwig (1959), in speaking with people indigenous to the area, learned that birth in the chacma baboon (*P. ursinus*) usually occurred under a bush while conspecifics watched. Bolwig, however, never witnessed a birth.

DeVore's (1963) comments on birth in wild olive and yellow baboons (*P. anubis* and *P. cynocephalus*) were restricted to those regarding immediate postpartum behavior. The mother was surrounded by other baboons soon after giving birth. The composition of the observers by sex was not specified, although DeVore mentioned that older juveniles and subadult females appeared to be most interested, while juvenile and young adult males expressed only a passing interest in the newborn. Older adult males, on the other hand, were seen to approach and touch the infant.

Vervet monkey. Although parturition was not observed by Gartlan (1969) in his field study of vervet monkeys (*Cerocopithecus aethiops*), prepartum and postpartum behavior were described. Adult females about to give birth did not isolate themselves from the rest of the group; and, after delivery, the females were often seen in close proximity to other animals. There was, however, a characteristic lack of interest in newborns by adult males.

Langurs. According to Jay (1963), the birth of a newborn free-ranging langur (*Presbytis entellus*) was a major point of interest for all adult and subadult females in the troop. Attention was centered upon the mother-infant dyad as soon as the newborn's presence was noted. The mother made no attempts to discourage curiosity so long as action by the observers was restricted to touching, smelling, and licking the neonate. Adult females were allowed to hold the infant within several hours of birth. In contrast, langur males were indifferent to newborns.

Poirier (1970) reported that juveniles and subadults (*P. johnii*) of both sexes were drawn to maternal groups; however, the attraction was presumably to the mother, not to the infant. The author interpreted such behavior as mediated by kinship. Adult males showed little interest in newborns but occasionally fulfilled a protective role.

Badham (1967) reported that, within hours after the birth of a leaf monkey (*P. obscurus*), the neonate was passed between females. Similarly, Hill (1972) witnessed the possession of a newborn douc langur (*Pygathrix nemaeus*) within hours of birth by a female who was not the mother.

Macaques. Gouzoules (1974) provided one of the few concerted efforts to study group behavior in response to parturition. Working with a captive group of twenty adult, subadult, juvenile, and infant stumptail macaques (*Macaca arctoides*), the birth of an infant to the highest ranking female was described in detail.

During early stages of labor two juvenile males showed evidence of sexual arousal in response to the pregnant female. The alpha male attempted to copulate with the female as the infant's head crowned. The female withdrew from the male with fear screams.

Placentophagia was restricted to the mother, although both a subadult and a young female showed interest in the placenta. Approaches by females were actively avoided by the mother, although two hours after birth the alpha male participated in a grooming bout with the mother and remained near her for almost two hours. Adult females showed the least interest in the delivery and postpartum situation. The individual who was most consistently associated with the mother after parturition was her juvenile son.

According to Rosenblum and Kaufman (1967), the interest of the members of two captive groups of pigtail (*M. nemestrina*) and bonnet (*M. radiata*) macaques in the discovery of a newborn was considerable. However, pigtail mothers appeared uneasy with the attention while bonnet mothers permitted a great deal of freedom with the newborn. Parturition itself was not reported.

Itani (1959) discovered that the invididuals of a wild Japanese macaque group (*M. fuscata*) usually maintained their distance from a newborn and its mother. The age range of "newborns" was not specified by the author.

Tinklepaugh and Hartman (1930) described the parturition of a female rhesus (*M. mulatta*) who shared a cage with another pregnant adult female. During delivery the latter monkey showed no special interest in the mother or neonate; indeed, the animal slept through most of the process, stirring only twice to stare at the baby or groom the mother briefly. In another reported situation, a male yearling aided its mother in licking the afterbirth from its newborn sibling.

Rowell et al. (1964) observed two live births in captive rhesus monkeys. In six other cases observations were begun within the first few hours after birth. In both of the former cases the mothers were closely attended by another adult female during the final stages of labor. These "aunts" were childless at the time. Childless females were twice observed to arch their backs and finger their own vulvas at the sight of an emerging infant. Likewise, final stages of birth elicited sexual excitement in several males, and mountings were reported. Placentophagia was restricted to the mother, although other animals appeared to be curious about the afterbirth. Only nulliparous females, however, were seen to taste the excretions. None of the cagemates was allowed to touch the newborn although favored individuals were sometimes allowed to visually scrutinize the neonates.

Mitchell and Brandt (1975) reported the reactions of various cagemates to the rhesus birth process. An adult female displayed grooming, lipsmacking, intense curiosity, and protective responses when placed with a birthing female. On at least two occasions adult females attempted to, or succeeded in, kidnapping the newborn.

Two births were witnessed in which an adult male was housed with a female in labor. The females' responses to the males were characterized by nervousness and withdrawal. The males, also nervous, nevertheless attempted to groom and inspect the females. Male curiosity in the afterbirth was noted, but placentophagia by the males did not occur. Both males were protective of the mothers and infants, as evidenced by threats directed at human observers (see Plates 7 and 8a-b). The authors suggest that, in the wild, male rhesus do not attend females at parturition. This assumption is supported by the females' avoidance of the males during parturition in the laboratory. Furthermore, Lindburg (1971) saw no evidence that free-ranging male rhesus pay any attention to the female during labor.

Colobus monkeys. Hill (1972) reported "infant sharing" in captive king colobus monkeys (*Colobus polykomos*) soon after birth. On at least two occasions the animals seen holding the neonates were siblings of the newborn. All animals except adult males participated in sharing. Wooldridge (1971) observed the birth of a captive black and white colobus monkey (*Colobus guereza*). Out of the seven conspecifics present during labor and parturition, only an adult female and a juvenile male showed any interest. However, the females in the group were seen to handle and care for the infant within hours of birth. In general, the mother did not object to the infant sharing.

Patas monkeys. Hall and Mayer (1967) witnessed postpartum behavior in a captive group of patas (*Erythrocebus patas*). The adult male showed no interest in the newborn, whereas the other adult female in the group was greatly attentive. The mother did not allow the female to touch her infant although this same female later became the infant's "aunt."

Apes and Humans

Hylobatidae. Chivers and Chivers (1975) reported that a free-ranging female siamang (*Symphalangus syndactylus*) isolated herself from her group (consisting of her mate, a subadult male, and a juvenile male) before birth but rejoined her groupmates within twenty-four hours of parturition.

Pongidae. Chaffee (1967) commented on two captive orangutan (*Pongo pygmaeus*) births from the same breeding pair. The male was present throughout the deliveries but was not reported to interfere in the process. Ullrich (1970), however, reported what is perhaps the most fascinating of all birth-related phenomena available in the literature. In this account of captive orangutans, the author reports that an adult male, the mate of a pregnant female, became sexually aroused during the latter's labor and attempted numerous times to copulate with her. Upon the crowning of the infant's head, however, the male's behavior changed drastically. Assuming a calm and non-sexual demeanor, the animal positioned himself at the female's raised perineum. He then put his mouth over the emerging newborn's head and pulled gently. Once the head was fully freed, the male used his hands to deliver the rest of the neonate's body from the birth canal. He held the infant until the female had righted herself; at that time he relinquished the newborn to its mother. Behavior virtually identical to this was observed once again in the same animals during a later birth. Ullrich speculates that such male assistance in parturition may have arisen as a result of the orangutan's arboreal life style; an infant delivered off the ground could easily fall to its death upon expulsion from the mother. The male's initial sexual attraction to the laboring female would serve to establish the close proximity necessary for aid giving. The crowning of the infant's head may then serve as an eliciting stimulus for the appropriate delivery behavior.

The male's "oral infant contact" behavior is particularly interesting in light of data obtained during the births of captive chimpanzees (*Pan troglodytes*). Fox (1929) reported he twice observed female chimps placing their mouths over the mouths of newly emerged infants and blowing conspicuously. Lemmon (1968) reported a similar event in captive chimps. The question is raised as to whether or not the mothers were artificially respirating their infants at these times. Further data are necessary to answer this question.

Nissen and Yerkes (1943) mention only one case in which an infant was found in the protective care of a young adult female. The mother was apparently oblivious to the infant.

Hominidae. Several recent studies have suggested that the human father's role in the birth process can be of considerable consequence. Tanzer (1973) reported that the husband's presence was necessary for the woman to report an ultimate or "peak" experience associated with birth. Similarly, women giving birth felt safer with their husbands present at delivery (Newton, 1973). Mazur and Warren (personal communication), in studying the effects of the Lamaze birth method, reported that the single most helpful technique (in terms of reduced pain) was having the husband present with his wife during labor. Further, immediately after birth the father can play a more active role in postpartum infant interaction than customary (Parke and O'Leary, 1975).

DISCUSSION AND SUMMARY

The most apparent feature in the literature is that the marmoset exhibited more interest in parturition than did adult males of other species. This undoubtedly relates to the monogamous social system of the marmoset. From a polyphyletic viewpoint, monogamous males typically play a relatively substantial role in the care of the offspring. The male marmoset is no exception. Because his investment in each offspring is comparatively high, great interest in the infant is to be expected. Conversely, the polygynous male squirrel monkey, which apparently has little contact with females in the wild and exhibits no infant-directed care, is reported to actively "retreat" from the vicinity of a birth. Similarly, langur males (*P. entellus* and *P. obscurus*) are reported to be particularly indifferent to infants. Notably, however, langur females are greatly interested in neonates and often assume an immediate role in infant-directed care. Male and female patas monkeys exhibit similar patterns of responding. The purpose of such female behavior is unknown, although kin selection, playmothering, and alliance formation have been suggested as etiological factors (Hrdy, 1976). Further, such female care may be particularly beneficial in social systems where the males play little or no active role in protection.

A characteristic interest in birth on the part of both male and female juveniles is noted in most species. For example, when placentophagia is seen in animals other than the mother, the participants are typically juveniles or subadults.*

Adult males are occasionally reported to show sexual excitement in the vicinity of laboring females. Such reports are limited to studies of captive animals in which the male and female are in close proximity. Thus, the male's excitement may be an artifact of the physical situation.

Increased research in the area of group response to birth in primates is greatly needed in order to clarify and extend the available information. Not only is such information inherently interesting, but also the application of theoretical ideas regarding primate behavior is potentially testable in the context of behavior related to birth.

REFERENCES

Badham, M. A note on breeding the spectacled leaf monkey at Twycross Zoo. *Int. Zoo Yearbook* **7**: 89 (1967).

Basilewsky, G. Keeping and breeding Madagascan lemurs in captivity. *Int. Zoo Yearbook* **5**: 132–137 (1965).

Bolwig, N. A study of the behavior of the chacma baboon, *Papio ursinus. Behaviour* **14**: 136–163 (1959).

Bowden, D., Winter, P., and Ploog, D. Pregnancy and delivery behavior in the squirrel monkey and other animals. *Folia Primatol.* **5**: 1–42 (1967).

Carpenter, C. R. The Howlers of Barro Colorado Island. In I. DeVore (Ed.) *Primate Behavior: Field Studies of Monkeys and Apes.* New York: Holt, Reinhart and Winston, 1965, pp. 250–291.

Carpenter, C. R. A field study of the behavior and social relations of howling monkeys (*Alouatta palliata*). *Comp. Psychol. Monogr.* **10**(48): 1–168 (1934).

Chaffee, P. S. A note on the breeding of orangutans. *Int. Zoo Yearbook* **7**: 94–95 (1967).

Chivers, D. J. and Chivers, S. T. Events preceding and following the birth of a wild siamang. *Primates* **16**: 227–230 (1975).

DeVore, I. Mother-infant relations in free-ranging baboons. In H. L. Rheingold (Ed.) *Maternal Behavior in Mammals.* New York: Wiley, 1963, pp. 305–355.

Doyle, G. A., Pelletier, A., and Bekker, T. Courtship, mating and parturition in

*Terry Maple has informed us that an adult male orangutan ingested, but later regurgitated, a placenta.

the Lesser Bushbaby (*Galago senegalensis moholi*) under semi-natural conditions. *Folia Primatol.* **7**: 169–197 (1967).
Dunbar, R. I. M. and Dunbar, P. Behavior related to birth in wild gelada baboons (*Theropithecus gelada*). *Behaviour* **50**: 185–191 (1974).
Fox, H. The birth of two anthropoid apes. *J. Mammal.* **10**: 37–51 (1929).
Gartlan, J. S. Sexual and maternal behavior of the vervet monkey, *Cercopithecus aethiops*. *J. Reprod. Fertil., Suppl.* **6**: 137–150 (1969).
Gouzoules, H. T. Group responses to parturition in *Macaca arctoides*. *Primates* **15**: 287–292 (1974).
Hall, K. R. L. and Mayer, B. Social interactions in a group of captive patas monkeys (*Erythrocebus patas*). *Folia Primatol.* **5**: 213–236 (1967).
Hanif, M. Notes on breeding the white-headed saki monkey (*Pithecia pithecia*) at Georgetown Zoo. *Int. Zoo Yearbook* **7**: 81–82 (1967).
Hill, C. A. Infant sharing in the family colobidae emphasizing *Pygathrix*. *Primates* **13**: 195–200 (1972).
Hopf, S. Notes on pregnancy, delivery, and infant survival in captive squirrel monkeys. *Primates* **8**: 323–332 (1967).
Hrdy, S. B. Care and exploitation of nonhuman primate infants by conspecifics other than the mother. In E. Shaw and R. A. Hinde (Eds.) *Advances in the Study of Behavior*, Vol. 6. New York: Academic Press, 1976, pp. 101–158.
Itani, J. Paternal care in the wild Japanese monkey. *Primates* **2**: 61–93 (1959).
Jay, P. Mother-infant relations in langurs. In H. L. Rheingold (Ed.) *Maternal Behavior in Mammals.* New York: Wiley, 1963, 282–304.
Kummer, H. Social organization of Hamadryas baboons. *Bibl. Primatol.* **6**: 1–189 (1968).
Langford, J. B. Breeding behavior of *Hapale jacchus* (common marmoset). *South African J. Sci.* **59**: 299–300 (1963).
Lemmon, W. B. Delivery and maternal behavior in captive reared primiparous chimpanzees, *Pan troglodytes.* Paper presented at the American Association for the Advancement of Science Meeting, Dallas, 1968.
Lindburg, D. G. The rhesus monkey in North India: An ecological and behavioral study. In L. A. Rosenblum (Ed.) *Primate Behavior: Developments in Field and Laboratory Research*, Vol. 2. New York: Academic Press, 1971, pp. 1–106.
Lucas, N. S., Hume, E. M., and Smith, H. H. The breeding of the common marmoset in captivity. *Proc. Zool. Soc. London* **107**: 205–211 (1937).
Mitchell, G. and Brandt, E. M. Behavior of the female rhesus monkey during birth. In G. H. Bourne (Ed.) *The Rhesus Monkey*, Vol. 2. New York: Academic Press, 1975, pp. 232–245.
Neville, M. K. Social relations within troops of red howler monkeys (*Alouatta seniculus*). *Folia Primatol.* **18**: 47–77 (1972).
Newton, N. Trebly sensuous woman. In C. Tavris (Ed.) *The Female Experience.* New York: Ziff-Davis, 1973, pp. 22–26.
Nissen, H. W. and Yerkes, R. M. Reproduction in the chimpanzee: Report on 49 births. *Anat. Rec.* **86**: 567–578 (1943).

Parke, R. D. and O'Leary, S. Father-mother-infant interaction in the newborn period: Some findings, some observations, and some unresolved issues. In K. Riegel and J. Meàcham (Eds.) *The Developing Individual in a Changing World*, Vol. 2. *Social and Environmental Issues.* The Hague: Mouton, 1975, pp. 653–663.

Poirier, F. E. The Nilgiri langur (*Presbytis johnii*) of South India. In L. A. Rosenblum (Ed.) *Primate Behavior: Developements in Field and Laboratory Research*, Vol. 1. New York: Academic Press, 1970, pp. 281–383.

Rosenblum, L. A. and Kaufman, I. C. Laboratory observations of early mother-infant relations in pigtail and bonnet macaques. In S. Altmann (Ed.) *Social Communication Among Primates.* Chicago: University of Chicago Press, 1967, pp. 33–42.

Rothe, R. H. Influence of newborn marmosets' (*Callithrix jacchus*) behavior on expression and efficiency of maternal and parental care. In S. Condo, M. Kawai, and A. Ehara (Eds.) *Contemporary Primatology: Proceedings of the Fifth International Congress of Primatology, Nagoya, Japan, 1974.* Basel: S. Karger, 1975, pp. 315–320.

Rowell, T. E., Hinde, R. A., and Spencer-Booth, Y. Aunt-infant interaction in captive rhesus monkeys. *Anim. Behav.* **12**: 219–226 (1964).

Sorenson, M. W. Behavior of tree shrews. In L. A. Rosenblum (Ed.) *Primate Behavior: Developments in Field and Laboratory Research*, Vol. 1. New York: Academic Press, 1970, pp. 141–193.

Sorenson, M. W. and Conaway, C. H. Observations on the social behavior of tree shrews in captivity. *Folia Primatol.* **4**: 124–145 (1966).

Stevenson, M. F. Birth and perinatal behavior in family groups of the common marmoset (*Callithrix jacchus jacchus*), compared to other primates. *J. Hum. Evol.* **5**: 365–381 (1976).

Tanzer, D. Natural childbirth: Pain or peak experience. In C. Tavris (Ed.) *The Female Experience.* New York: Ziff-Davis, 1973, pp. 26–33.

Tinklepaugh, D. L. and Hartman, C. G. Behavioral aspects of parturition in the monkey. *J. Comp. Psychol.* **11**: 63–98 (1930).

Ullrich, W. Geburt und natürliche Geburtshilfe beim Orang-Utan. *Der Zool. Garten* **39**: 284–289 (1970).

Williams, L. Breeding Humboldt's wooly monkey (*Lagothrix lagothricha*) at Murrayton Wooly Monkey Sanctuary. *Int. Zoo Yearbook* **7**: 86–89 (1967).

Wooldridge, F. L. *Colobus guereza*: Birth and infant development in captivity. *Anim. Behav.* **19**: 481–485 (1971).

5
Baboon Behavior under Crowded Conditions*

Robert H. Elton

Primate Research Program and Department of Psychology
Eastern Washington University
Cheney, Wash.

INTRODUCTION

> Under some circumstances crowding may have disastrous effects on rats, mice, rabbits, and other animals, but crowding does not have generally negative effects on humans. People who live under crowded conditions do not suffer from being crowded. Other things being equal, they are no worse off than other people.
>
> J. L. Freedman, 1975, P. 1

Many areas of the behavioral sciences have problems with equivocal data, but in population density or crowding studies, the line of demarcation seems to be whether the subjects of the studies were human or nonhuman. With few exceptions, animal studies, chronic or acute, show powerful effects. Also, with few exceptions, human studies show weak or negligible effects. Consequently, interpretation is difficult for theorists in this area of study. Altman (1975), for example, has argued that if psychological stress were the result of increased density, human subjects in crowding experiments might have engaged in coping behaviors that alleviated the stress, and consequently little or no effect would have been detectable. Often human

---

*This study was supported by National Institutes of Health grant RR00166 to the Regional Primate Research Center at the University of Washington, and by a grant from the Primate Research Program, Eastern Washington University. I wish to thank the following students who served as observers during the study: Steve Hollinsworth, Becky Cooper, Mary Bly, William Baker, and Sally McCallum. I also thank Dr. Gerald Blakley and the staff at the Medical Lake Field Station of the Regional Primate Research Center at the University of Washington. Brian Anderson, who followed the study from conception to completion, and Phyllis Wood, who did the art work for the manuscript, deserve very special thanks.

subjects have known the duration of the crowding manipulation; this awareness might have mediated dissipation of the stress effect. Consequently, Altman has suggested that stress measures be taken throughout experiments rather than at the end, to allow detection of adaptation to the situation. Apparently humans have mechanisms that attenuate or even dissipate stressful outcomes when sufficient information is available. Nonhuman animals may not have these mechanisms or may have insufficient information concerning temporal or other limiting factors that would allow cognitive dissipation of density effects. Thus, the response of nonhuman animals might be reflected in social, physiological, or individual indices.

Nonhuman Primate Studies of Crowding Effects

Nonhuman primates may serve as valuable models for studying density stress effects because of their physiological similarities to human primates. Such studies may also provide valuable information which will lead us to better understand density-related dangers to natural populations of primates and lead to more humane and effective treatment of laboratory and zoo primate groups. Although the now classic experiments by Calhoun (1962), which produced severe pathology and social disintegration in overcrowded rat populations have stimulated considerable research and speculation, few studies have investigated crowding in nonhuman primates.

Urban versus Rural Monkeys. Singh (1968) investigated the social behavior of rural and urban monkeys. Two sex-balanced groups of rhesus monkeys, six caught in city areas and six from forest areas of India, were tested in three situations: dominance tests; social behavior tests; and group social interaction tests. The urban animals were the most aggressive and dominant in all tests. Singh suggested that the highly aggressive nature of the urban monkeys was essentially due to competition for survival in the crowded environment in which they lived prior to the study.

Effects of Environmental Manipulation. Southwick (1967) studied twenty-five feral rhesus monkeys in situations of varied complexity and available space. In one study a concrete pool was filled with water to create a new activity focus and water source, with the expectation that a pool of water might reduce the level of agonistic behavior by providing a new outlet for activity. In a second experiment the cage was partitioned midway across by a wire fence extending from floor to roof with a small open doorway. In a third experiment all monkeys were forced into one half of the cage and the partition door was closed. In the first two experiments minor increases in the frequency of threats, submissive responses, and attacks occurred. Crowding (the third experiment) seemed to exert a major influence on the form and frequency of agonistic behaviors, including increased attacks and fights (but see Erwin, this volume). Interestingly, social changes (for example, the introduction of a stranger to the group) produced much more dramatic increases in aggressive interactions than did alterations in the physical environment.

Successive Short-Term Crowding Effects. Effects of acute, short-term crowding were studied in a troop of Japanese macaques that were kept in a two-acre corral (Alexander and Roth, 1971). Crowding was accomplished by confining the animals in an adjacent pen which restricted the troop to 2.3 percent of their accustomed area. A tunnel connecting the pen and corral permitted transfer of the troop so that periods in their large environment could be alternated with experimental periods in the confined area. There were three experimental periods, each lasting four to six days. Crowding produced sharp increases in aggression among males which disappeared when the troop returned to the corral. Mobbing occurred in the large unfamiliar environment prior to the experiment, and in the small familiar environment during the experiment. Alexander and Roth (1971) hypothesized that removal from a familiar habitat is sufficient to provoke mobbing independent of habitat characteristics.

Physiological Effects of Crowding. The effect of crowding on primate physiological systems has also received little attention. A frequently observed correlate of population density (particularly in rodents) is increased adrenal size and weight (Christian, 1961). The growth, body weight, and reproduction of tree shrews (*Tupaia glis*) is detrimentally affected by crowding (Autrum and von Holst, 1968). Mixed sex groups of subadult rhesus monkeys, maintained for two years under chronic crowding, exhibited a relationship between behavior and ACTH-response level, a stress measure (Sassenrath, 1970). The slow time course of ACTH-response changes suggests that stress-induced changes in certain physiological systems can be effected and maintained only when stress levels are constant, continuous, and sustained over periods of time as long as weeks or months. Sassenrath's results suggest that research on crowding effects should be continued across extended periods to allow the full course of stress-response development.

AN EXPERIMENTAL STUDY OF CROWDING

Method

Subjects. The subjects for this study were thirteen baboons (*Papio anubis*); age, sex, and rank distribution are summarized in Table 5-1.

Table 5-1. Sex, Age, Rank, Classification of the Troop

Individual Identification and Rank	Center ID Number	Age Class	Date Arrived[a] or Date Born[b]	Sex	Introduction (kg)	Estimated Age at End of Experiment
A[a]	6760	Adult	1967	Male	29.6	7 years
A[a]	67285	Adult	1967	Female	14.9	6 years
B[a]	67235	Adult	1967	Female	16.2	6 years
C[a]	67132	Adult	1967	Female	18.6	6 years
D[a]	6868	Adult	1968	Female	17.6	6 years
A[b]	M69121	Juvenile	10/13/69	Female	5.8	3 years
B[b]	M69132	Juvenile	10/24/69	Female	d	3 years
C[b]	M705	Juvenile	1/3/70	Female	d	$2\frac{1}{2}$ years
D[b]	M71226	Juvenile	9/9/71	Male	d	$1\frac{1}{2}$ years
E[b]	M71273	Juvenile	11/17/71	Male	d	$1\frac{1}{4}$ years
F[b]	C	Juvenile	6/24/72	Male	d	1 year
G[b]	C	Juvenile	9/9/72	Female	d	$\frac{1}{2}$ year
H[b]	C	Juvenile	12/16/72	Female		$\frac{1}{4}$ year

[a]Wild captive.
[b]Captive born.
[c]No recorded number.
[d]No recorded weight.

Apparatus. The experimental area was a windowless concrete room, 4.79 m long X 3.35 m wide X 3.04 m high. Inside the room were two steel pipe seats attached to the back and sidewall, a drinking fountain, a feeding trough, and a ventilation fan. A 2.93 X 3.35-m movable wall, constructed of sheet metal and angle iron, was placed at one end of the cage. A one-way observation mirror (4.9 X 11.9 cm) was centered on the wall 1.22 m up from the bottom. A sheet of clear plexiglas was placed over the one-way mirror to prevent breakage.

Feeding and Cleaning. The animals were fed once in the morning and once in the afternoon. The floor of the room was covered with sawdust chips to enable the animals to simulate foraging. During wall movements and cage cleaning the animals were moved into a small adjacent room.

Observations. Observers were students trained to identify individual animals and to classify behaviors in accordance with the categories used in this study. They vocally recorded all behavioral observations with portable cassette tape recorders. Data from the thirty-minute recordings were transcribed onto data sheets within an hour of observation. During reliability tests, headphones with instrumental music played over them were used to eliminate verbal cues, while a senior investigator and an observer conducted simultaneous observations. Reliability tests were scored by dividing the number of agreements plus disagreements into the number of agreements times 100. Observations were made at least once a day, five days per week. Each observation consisted of watching six randomly chosen animals for five minutes each.

Experimental Manipulations. The animals were moved into the experimental cage and allowed to adapt to the new habitat and each other for six months. The experiment was divided into five periods, with the number of days per time period differing slightly so that wall movements never occurred on a weekend when animals could not be observed. Period one lasted forty days and was used for baseline (established normal frequencies of behavioral categories). The second period, also forty days long, initiated the first crowding pe-

Fig. 5-1. A schematic representation of the baboon group prior to any wall movement.

riod. The third, fourth, and fifth periods were forty-three, forty-one, and thirty days, respectively. At the beginning of each crowding period the subjects were moved into a small adjacent room while the wall was moved inward 2 ft, thus decreasing available floor space by 2.04 m. The fifth period reduced the available space to 50 percent of the original area.

Results

The twenty-nine behaviors constituting the four categories used in this experiment are published elsewhere (Elton and Anderson, 1977); however, the four categories of concern for this report are social behaviors, agonistic behaviors, sexual behaviors, and individual behaviors. The number of observations per day per animal varied due to the subject selection technique and to the different number of observers on any one day. Data from only one observer were used for each day's analysis even if more data were available, because there were days when only one observer was present. On days with more than one observer, the observation used was randomly selected; however, data from any observer whose reliability fell below 85 percent were discarded. The troop was grouped for analysis into adults and juveniles.

Social Behaviors. The mean frequency of social behaviors for the group overall, adults, and juveniles is plotted in Fig. 5-2. Column I is the baseline period. For both adults and juveniles there was an initial increase in social behavior following the first wall movement (Column II). After the last wall movement there was a decrease reaching baseline levels for the overall and adult animals, even though the space available to the group had been decreased by 50 percent thus forcing them into a situation increasing the likelihood of social interaction.

A representative behavioral category was contact. We defined contact as any social behavior or contact not otherwise defined by groom, social investigate, embrace, mouth-mouth, play, huddle, or proximity. The frequency of contact that occurred during the study is presented in Fig. 5-3. This category exhibited a clear difference between adults and juveniles as much of this category was comprised of play activity and the adult animals seldom played. The general pattern exhibited by adults was also apparent in the juveniles' data.

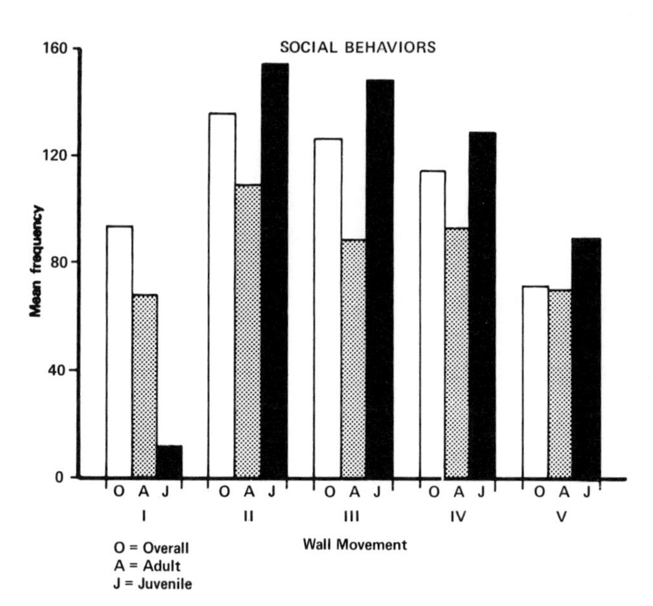

Fig. 5-2. The mean frequency of social behaviors for adult, juvenile, and the group overall for baseline (Column 1) and each wall movement.

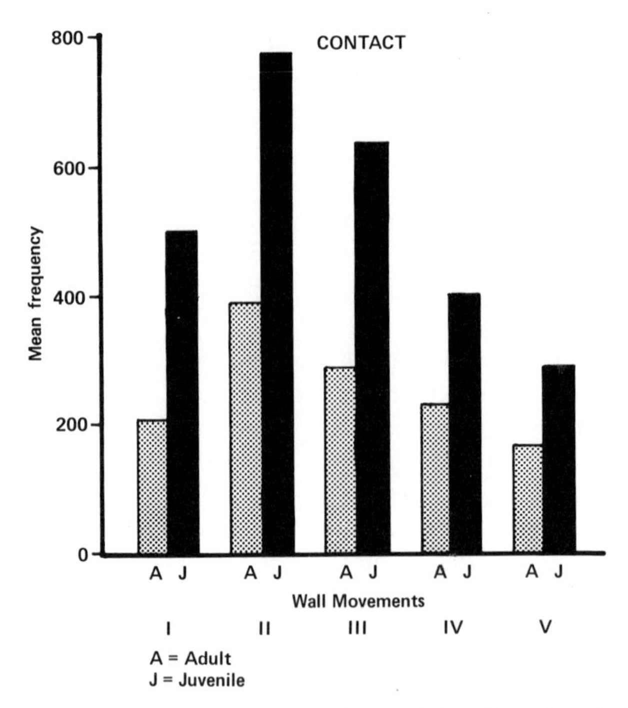

Fig. 5-3. The mean frequency of social contact by the adult and juvenile animals during the study.

Agonistic Behaviors. Figure 5-4 presents the mean agonistic behaviors recorded for the group. There was an increase in aggression for both juveniles and adults following the first wall movement and a continued increase with each additional wall move until the fifth wall movement when there was a decrease.

Sexual Behaviors. Sexual behaviors generally followed a pattern similar to that of the agonistic behaviors. There was an increase in behavioral frequencies as available space decreased. Figure 5-5 illustrates this behavior across wall movements.

The mean frequency of each category of sexual behavior during each wall movement phase is presented in Table 5-2. For adults sexual presents were more frequent than any other category of sexual behavior, probably because this posture is also often used as an appeasement gesture (in addition to its sexual function). Sexual behav-

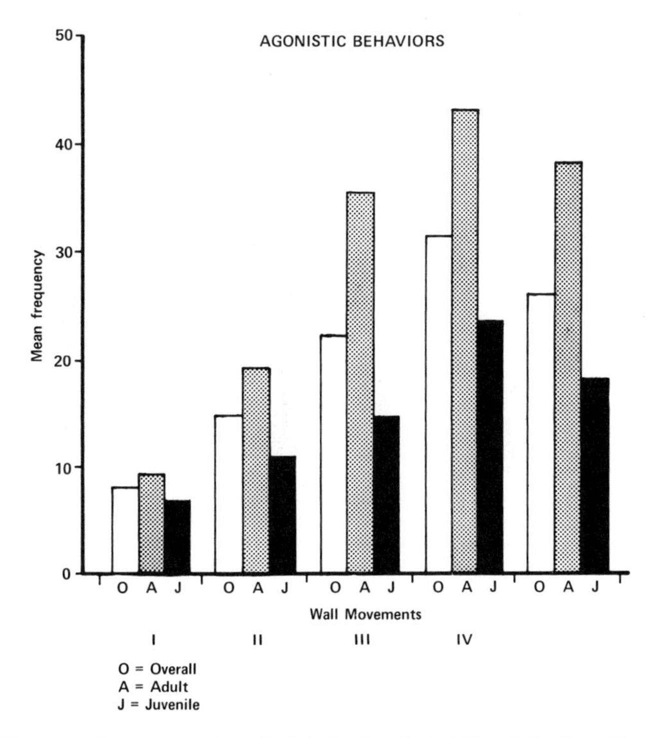

Fig. 5-4. The mean frequency of agonistic behavior plotted for adults, juveniles, and overall for each wall movement.

iors emitted by the juveniles were difficult to interpret due to maturational differences; however, they followed the pattern of the adult behaviors rather closely. Groups of Anubis baboons in the wild seldom contain a single adult male and the adult male of the studied group was a young adult; both of these factors might have influenced the frequency of behaviors in this general category.

Individual Behaviors. The category "individual behaviors" served as a category for behaviors of a nonsocial nature, or those that could not be categorized elsewhere, including drink, eat, forage, rest, automanipulate, vocalize, self-play, travel, and passive. The frequency of these behaviors is presented in Fig. 5-6 for the adults, juveniles, and total for each wall-movement phase.

In summary, as the available space decreased, the animals' behavior

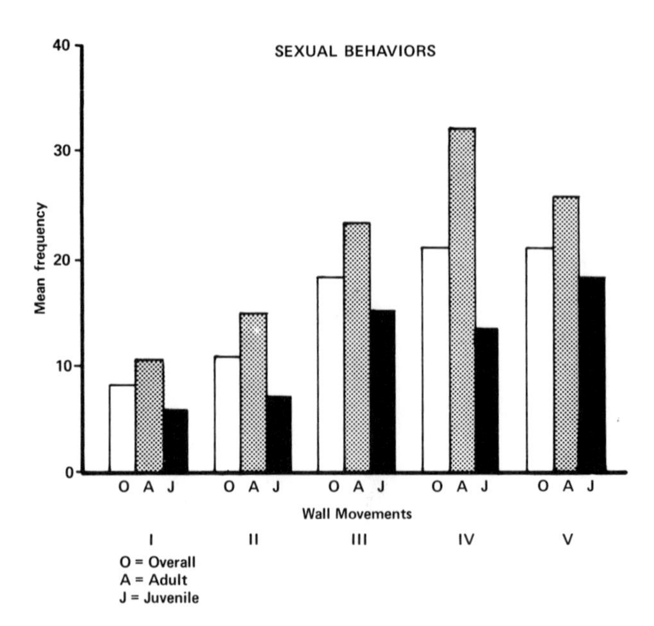

Fig. 5-5. Sexual behaviors plotted over wall movements for adults, juveniles, and overall for each wall movement.

Table 5-2. Mean Behaviors Recorded Comprising Sexual Category

Adults	Wall Movement				
	I	II	III	IV	V
Sex present	48	63	87	136	115
Mount	4	8	26	18	14
Masturbate*	3	3	4	3	1
Juveniles	**Wall Movement**				
	I	**II**	**III**	**IV**	**V**
Sex present	27	33	52	54	53
Mount	17	24	59	45	72
Masturbate†	3	2	1	5	8

*all Alpha male
†all Male Juveniles (D, E and F)

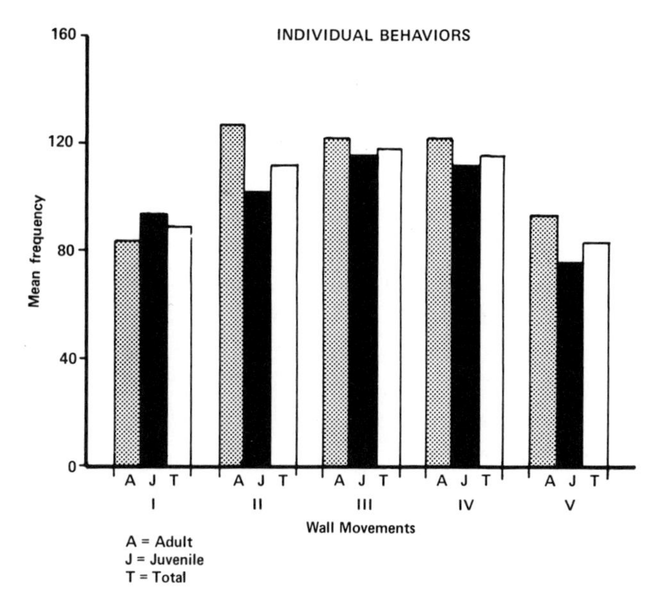

Fig. 5-6. Individual behaviors plotted over wall movements for adults, juveniles, and overall.

decreased, and the behaviors that did occur were primarily sexual or agonistic, rather than affiliative, social behaviors.

Discussion

In this experiment, crowding was accomplished by artificially imposing spatial restrictions on a captive group of baboons. The result of crowding was an increase in intratroop agonistic and sexual behaviors. The expressed aggression was hierarchically organized in essentially the same way as it was when the troop was less crowded. Although the dominance hierarchy in this troop remained extremely stable throughout the experiment, there was one exception; the lowest female in the hierarchy, Female D, had an infant and took over Female C's position in the hierarchy until Female C also gave birth. Due to the constantly decreasing available space, the animals appeared to enforce their positions in the hierarchy with increased threat, charge, and fighting behaviors.

Each wall movement caused what subjectively appeared to be stress or tension in the individual animals and the group as a whole. An interesting development was that the animals appeared always to adapt to the ever-increasing restrictions being placed on them by their environment. During the first two wall movements a similar pattern of behavior occurred among the animals. Approximately one and a half weeks after each of these wall movements, there was a noticeable increase in tension and general activity which continued until a few days before the end of the period. Then, as before, the animals appeared to adapt and returned to baseline levels of behavior. Of note is the fact that, during this fourth period, some behaviors appeared which had not been previously observed in this troop of baboons. Juvenile animals chewed repetitively on their own hands and feet, and licked the tongues and bit the teeth of other animals. An atypical behavior which appeared in some of the female juveniles was the pulling of hair out of other animals (by the handful) and eating it. As with the agonistic behaviors, these behaviors tended to drop out towards the end of the fourth period. Figure 5-7 is an artist's depiction of the group's behavior during the fourth wall movement. In

Fig. 5-7. The group's behavior following the fourth wall movement.

the beginning of the fifth period, there was again an almost immediate increase in agonistic behaviors. Also, within a week and a half the atypical behaviors among juvenile animals had not only reappeared, but had increased in frequency and intensity; for example, juveniles pulled hair out of adult females on the lower end of the hierarchy, and infants became socially withdrawn. During this period of social withdrawal, animals were observed engaging in numerous self-directed behaviors (self-play, self-groom, manipulate, etc.) which sometimes occurred for periods of an hour or more without interruption.

A few days after the start of the fifth period, a "riot" occurred. A piece of glass covering one of the two holes in the wall used for light sources was broken out by the monkeys, which resulted in a majority of the juveniles escaping through the hole to the observation side of the wall. A new glass plate was installed. The alpha male then broke it out and was found outside the wall on a weekend. He had broken the one-way mirror and had torn down most of the plastic covering the observation area. Another interesting occurrence during this

Fig. 5-8. The group's behavior following the fifth wall movement.

fifth period was an instance of mobbing. Juvenile A was mobbed by all four adult females for a period of about forty-five minutes, during which time the females took turns threatening, chasing, and biting this older juvenile female. Along with increases in agonistic behaviors to the extent of mobbing, we observed that reprimands had not only increased in frequency, but in intensity as well. At the start of the experiment, reprimands usually consisted of mild threats or arm or tail pulling. Towards the end of the experiment reprimands consisted of arm and leg stretching accompanied by numerous bites and vigorous threatening gestures. The alpha male was observed on numerous occasions throwing juveniles and infants across the cage area. The artist's concept of the troop following the last wall movement is presented in Fig. 5-8.

Social disintegration, as well as individual pathology, was the end result of crowding in this group of baboons. Disintegration, interestingly, first began with infants and juveniles, then with the females lowest in dominance, and finally was observed in the adult male of the group.

REFERENCES

Alexander, B. and Roth, E. The effects of acute crowding on aggressive behavior of Japanese monkeys. *Behaviour* **39**: 73–88 (1971).

Altman, I. *The Environment and Social Behavior.* Monterey, Calif.: Brooks Cole, 1975.

Autrum, H. and von Holst, D. Sozialer 'stress' bei Tupajas (*Tupaia glis*) und seine Wirkung auf Wachstum, Korpergewicht und Forthphlanzung. *Z. Vergleichende Physiol.* **58**: 347–355 (1968).

Calhoun, J. B. Population density and social pathology. *Sci. Am.* **206**: 139–148 (1962).

Christian, J. J. Phenomena associated with population density. *Proc. Nat. Acad. Sci.* **47**: 428–449 (1961).

Elton, R. H. and Anderson, B. V. The social behavior of a group of baboons (*Papio anubis*) under artificial crowding. *Primates* **18**: 225–234 (1977).

Freedman, J. L. *Crowding and Behavior.* San Francisco: Freeman, 1975.

Sassenrath, E. N. Increased adrenal responsiveness related to social stress in rhesus monkeys. *Hormones and Behavior* **1**: 283–298 (1970).

Singh, S. D. Social interactions between the rural and urban monkeys (*Macaca mulatta*). *Primates* **9**: 69–74 (1968).

Southwick, C. H. An experimental study of intragroup agonistic behavior in rhesus monkeys (*Macaca mulatta*). *Behaviour* **28**: 182–209 (1967).

6
Aggression in Captive Macaques: Interaction of Social and Spatial Factors*

J. Erwin

Department of Psychology
Humboldt State University
Arcata, California

INTRODUCTION

Fighting is a fairly common occurrence in primate groups even in natural settings, but trauma due to aggression is an especially pressing problem in captive groups of macaques and baboons. The intensification of destructive violence under captive conditions relative to free-ranging situations is well known (Nagel and Kummer, 1974; Rowell, 1967), but the causal factors are not very well understood.

A number of potential influences on aggressive behavior have been suggested (Nagel and Kummer, 1974), including the following: competition of males over females, conflict over infants, disturbance of social organization (e.g., dominance hierarchies), competition for food, competition for space (e.g., crowding and territorial disputes), absence of cover, reduction of utility of ecological functions (e.g., defense against predators), temporal distortions (e.g., reduced time required for foraging as a result of provisioning of food), incompatible artificial composition of groups and distortion of social roles.

*The research reported here was funded in part by each of the following USPHS/NIH grants: MH22253 to G. Mitchell; HD04335 to L. Chapman; HD06367-01 to W. A. Mason; RR00169 to the California Primate Research Center; RR00166 to the Regional Primate Research Center at University of Washington; and HD08633 to Gene P. Sackett. Innovative Education Awards from University of California, Davis, and University of Washington also contributed support to these projects. The manuscript was prepared with partial support from USPHS/NIH grants HD00973-12 and HD04510 to the Institute on Mental Retardation and Intellectual Development and the John F. Kennedy Center for Research on Education and Human Development. I am especially grateful to Nancy Erwin, who assisted in the conduct of the research and the preparation of the manuscript.

Some of the factors listed above have been evaluated experimentally by several investigators under a variety of conditions with the result that some high-risk conditions have been firmly established. In other cases, increased evidence has demonstrated the danger of premature conclusions and has pointed out the necessity of careful research on interactions between individual animals and their physical and social environments. In this chapter I will describe some research I have done (along with students and colleagues) at the California Primate Research Center and the University of Washington's Regional Primate Research Center Field Station, and will discuss the results in the context of the research of other investigators. The research I will describe here has derived from two sources, one a project designed to assess development of dyadic relationships (social attachments) in rhesus monkeys (see Mitchell et al., this volume), and a second project designed specifically to discover and eliminate the causes of bite wounding in a large domestic breeding colony.

SOCIAL BONDS AND AGGRESSION

One of the most striking features of primate social organization is the maintenance of group integrity. Most primates live in well-defined groups, and changing from one social group to another is surprisingly uncommon, particularly for females. For macaques, as well as many other primates, group cohesion appears to be based in part on specific long-term emotional bonds between individuals within groups.

Under normal social conditions the earliest bond is that established between the infant and its mother, but by the second half-year of life the amount of time spent in direct contact with the mother has decreased substantially, and interaction with peers has markedly increased. During the second and third years of life, relationships with peers supersede all others. These peer relationships are typically closest among individuals of the same gender.

Laboratory studies of relationships among juvenile and adolescent rhesus monkeys have documented the existence, specificity, and persistence of affectional bonds based on early peer interactions (see Mitchell et al., this volume). In a study which directly addressed this question, we reared like-sexed pairs of rhesus monkeys together throughout the second year of life. Their responses to brief separation and reunion were noted at about two years of age (Erwin et al., 1971), and they were separated again. Two years later each animal was again reunited with its familiar peer and responses to the familiar peer were contrasted with responses to an unfamiliar peer (Erwin et al., 1973).

In each case in which a female was paired with an unfamiliar peer, one member of the dyad viciously attacked the other; in each case an animal bit the other animal on the back in a very similar manner across all pairs. On the other hand, no female *ever* exhibited aggression of any sort toward the familiar animal with which she had spent the second year of her life. Despite separation for two full years, there was apparent recognition of familiar individuals and the demonstration of affection was especially touching in one pair of females. More than ten minutes had elapsed after the monkeys had been introduced into the test cage. Although they had spent considerable time looking at each other, they had avoided *mutual* gaze and had not established physical contact. As one female sat in the corner facing away from the other, the second animal approached gradually and touched her. Immediately, the first female turned and embraced the second, moaning (girning) and lipsmacking intensely. The two remained in contact, mutually grooming and embracing for the remainder of the test period. Other familiar females embraced, groomed, or maintained proximity to one another with no display of aggressive behavior.

While the picture was not as clear for males, there was some evidence that the familiarity dimension influenced aggression. Two unfamiliar males engaged in severe fighting, with multiple injuries to one within the first few minutes after they were paired. All familiar males immediately established contact and embraced each other, but in two such pairs what appeared to be play biting erupted into brief fights. The other familiar pair of males immediately established contact with mutual mounting (including anal penetration), and enduring bouts of mutual grooming. When the latter males were paired with unfamiliar males, no fighting ensued, but neither did any social contact occur. Because the one pair of unfamiliar males fought so severely that one was injured, the members of that pair were separated early in their test session, and the data base for males was reduced. Even so, frequency of locomotion (an indicator of disturbance) was significantly higher in unfamiliar males than in males reunited after a two-year separation.

When data were examined without regard to the sex of the subjects, some additional differences in behavior related to the familiarity-unfamiliarity dimension became apparent. All these differences confirmed the hypothesis that social bonds established with like-sexed peers during the second year of life persisted through the extensive two-year period of separation.

In a similar study, young adolescent rhesus monkeys were paired across sex for about six months (Erwin et al., 1973; Erwin and Mitchell, 1975). As in the previous study, the animals were separated and tested two years later with familiar and unfamiliar other-sexed peers (Erwin and Flett, 1974). The only overt aggression that occurred was perpetrated by males against unfamiliar females. No threatened aggression was directed toward a familiar peer by any male or female. Males and females both threatened the unfamiliar peers with which they were paired. Copulation was more frequent among familiar than unfamiliar pairs. These results indicate that heterosexual bonds formed in early adolescence can also persist over long periods of time, and that the absence of such bonds may influence patterns of aggression.

The social bonds that develop among members of free-ranging groups are probably the basis of group cohesion, and the xenophobic response (fear or hatred of strangers) is negatively correlated with the social bond. That is, the bond is specific, exclusive, and enduring; while many such bonds form between group members, they act to exclude intruders. The disruptive consequences of introduction of strangers into captive groups of macaques and baboons is well documented. In one well-known case (Hall, 1964), the introduction of an adult male and adult female baboon into a group of seventeen baboons at the Bloemfontein Zoo in South Africa resulted in utter chaos. The fighting that ensued devastated the group, with the majority of the animals being killed or dying later of the injuries sustained during the violent aggressive upheaval. Sadly, that scenario has been repeated many times in zoological exhibits and research colonies. During one of our studies at the University of Washington Field Station, a similar outburst was noted (Erwin et al., 1976). Two groups of pigtail macaques were merged after the adult male from

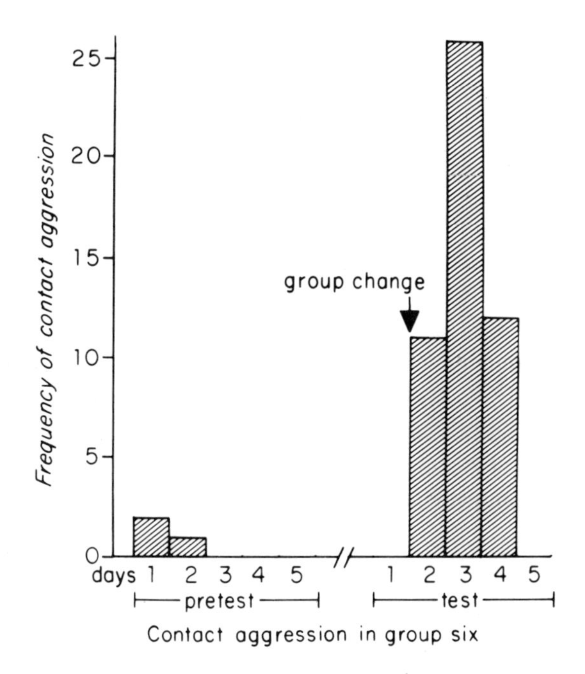

Fig. 6-1. Contact aggression increased dramatically when two groups of pigtail monkeys were merged.

one group had been removed. As Fig. 6-1 shows, the rate of contact aggression increased drastically following group change. During the period of heightened aggression, one female delivered a stillborn infant, and one female died of bite wounds within three days of the group merger. Over successive weeks nearly all the animals in the group sustained serious injuries and several additional animals died.

Introduction of unfamiliar animals to captive rhesus groups typically results in aggression and violence (Bernstein, 1964; Southwick, 1967; Fairbanks et al., 1977), and Southwick et al., (1974) have recently demonstrated the generality of this phenomenon to natural groups of rhesus in India. Unfamiliar individuals were introduced to wild groups of rhesus. Infants were accepted, even adopted, but all others were attacked and repelled. In a later section of this chapter, that dealing with formation of new groups in captivity, additional evidence of the importance of social bonds and alternative xenophobia in the production of violence in captive groups will be presented. Suffice it to say, at this point, that familiarity endures as a deterrent to violence and unfamiliarity contributes to the risk of violence.

CROWDING AND AGGRESSION

The supposition that crowding causes aggression is well accepted (Ardrey, 1966; Calhoun, 1961; Lorenz, 1966)—perhaps too well accepted. Bernstein and Gordon (1974) recently pointed out that experimental studies on crowding effects of nonhuman primate aggression have failed to produce more than temporary increases in intragroup aggression. There have not been many deliberate studies of relationships between available space and aggressive behavior, and the results are ambiguous at the present time. This is partly due to the lack of clear formulation of problems related to crowding.

What is Crowding?

Some investigators, particularly in the social sciences, have defined crowding in terms of subjective ratings about "feeling" crowded. Although this matter is of interest to us with regard to the improvement of housing for captive primates, in the sense that we would like to be able to find ways of evaluating negative feelings about crowding, it is essential that we first settle on definitions regarding the physical parameters of crowding, or more precisely, of *density*.

Density. Density refers to the average number of organisms per unit of defined space. Thus, density can be represented in the following manner:

$$d = N/S$$

where d is density, N is the number of organisms, and S is space. Space is frequently expressed in terms of area, such as square miles, acres, hectares, or square meters. For the description of captive quarters, the latter is usually the most appropriate. It is essential to recognize that length and width are not the only components of environments, and that vertical and temporal dimensions are also important, but let us first consider a more simplified case, that of spatial area. It is important to remember that for purposes of this discussion, density is expressed as the number of individuals per unit of space. Thus, the density for a group of ten baboons in a 400-m^2 enclosure is 0.025 baboons/m^2.

Individual Space. Another concept of the relationship between available space and the number of individuals within that space is frequently confused with or used interchangeably with the term density. It is actually the reciprocal of density as defined above. I shall call this *individual space*, that is, the average amount of space available to each individual. Individual space can be expressed in the following manner:

$$S_I = 1/d = S/N$$

where S_I is individual space, $1/d$ is the reciprocal of density, S is space (area), and N is the number of organisms. It should be noted that individual space does not define the *actual* amount of space available to each individual, but rather the *average* amount of space per individual. Thus, the individual space for a group of ten baboons in a 400-m^2 enclosure is 40 m^2/baboon. This obviously does not actually mean that each baboon is limited to 40 m^2; theoretically, of course, each animal would have access to all of the 400 m^2 not occupied by other group members. Captive caging requirements are often expressed in terms of individual space, but in some ways that will become apparent as this chapter progresses, the use of density is preferable to the use of individual space.

How Does Crowding Occur?

Crowding is actually a relativistic term. We become more crowded or less crowded relative to other situations. Changes in density can occur in several ways, the most obvious of which involve change in one or both of the components of density.

Spatial Alteration of Density. Density varies as a direct function of changes in the amount of available space, and this can occur independently of alteration of the number of individuals present in a particular space. For example, a group containing ten baboons may be temporarily moved from their 400-m^2 enclosure to a small pen with an area of only 40 m^2. The resultant crowding would be expressed as an *increase* in density from 0.025 baboons/m^2 to 0.250 baboons/m^2 and a decrease in individual space from 40 m^2/baboon to 4 m^2/ baboon. Thus, the relationship between crowding and density is direct and between crowding and individual space is inverse. The change in density resulting from spatial change can be represented as:

$$d = N/\Delta S$$

Experimental studies of crowding employing manipulation of space while holding N constant are called studies of the effects of changes in *spatial density.*

Social Alteration of Density. Density also varies as a direct function of changes in the number of individuals within a specific space. For any particular space, N may vary due to reproduction, mortality, or artificial addition or withdrawal of animals from groups. For example, we could add ninety baboons to our group of ten baboons in the 400-m^2 enclosure. This would increase the density from 0.025 baboons/m^2 to 0.25 baboons/m^2 and would decrease individual space from 40 m^2/baboon to 4 m^2/baboon. Remarkably, the density figures for this major manipulation of N results in the same density and individual space values as the example of alteration of spatial density above. Yet all the change in density is accounted for by variation in the number of individuals in the enclosure as expressed in the following manner:

$$d = \Delta N/S$$

Experimental studies of crowding employing manipulation of the number of individuals within static spaces are known as studies of the effects of changing *social density.*

Social *and* Spatial Alteration of Density. Of course, it is possible for social and spatial density effects to be evaluated simultaneously by examining the differential effects of spatial change on groups of different sizes. This should be an especially valuable area for future research, because it seems likely that spatial change affects groups with many members much differently than it affects those with few. A model based on variance seems especially appropriate for the integration of these factors, e.g.,

$$\sigma_d^2 = \sigma_N^2 + \sigma_S^2 + \sigma_{SN}^2$$

Such a model would, of course, allow us to partition the relative amounts of variance accounted for by social or spatial factors.

Limitations on the Components of Density

The importance of carefully examining and defining the issues relevant to crowding is to be able to generate expected values against which the results of observations and experiments may be evaluated. I will briefly direct attention to some aspects of spatial and social density manipulations before further extending the elaboration of their uses in generating useful expected value predictions.

Spatial Density. An important aspect of the spatial-density experiment is the necessity to assure that the manipulation involves *only* manipulation of space. It is not sufficient to divide a pen in half and confine animals in one half of the pen if one side contains a "Jungle Jim" and the other contains a swimming pool. The manipulation, in such a case, involves qualitative confounds of spatial change. Of course, in the case of two equally sized areas that are qualitatively different, there are appropriate experimental designs involving repeated testing and reversal of conditions that would convert such a situation into a benefit by lending generality to the results *if* the results were sufficiently similar in the different situations. On the other hand, it is very difficult to disentangle the contributions of quantitative and qualitative spatial features in a situation in which animals are crowded from a large pen containing trees and large concrete pipes into a tiny cement and woven wire enclosure; regardless of repeated trials, the qualitative differences between the two environments remain confounded with the quantitative spatial manipulation. The best example of a study designed to exclusively manipulate quantity of space was that by Elton (this volume); although social changes in the group became confounded with spatial manipulation, the planned reversal phase of the study would have provided a valuable comparison had it been possible to complete that phase of the study.

Social Density. There are many more problems in assessing expectancies about variation in social density than those associated with spatial density. These, of course, arise from the active nature of organisms in contrast to the passive nature of spatial considerations. There are two general kinds of social-density studies, longitudinal and cross-sectional, which I will treat separately.

Longitudinal studies. Social-density changes within groups must be thought of as including the important dimension of *time*. The effects of temporal social manipulations or occurrences such as births, deaths, artificial removals from or additions to groups must be assessed relative to previous and subsequent base rates. Like the problems with spatial-density studies, longitudinal studies of this sort include many qualitative changes inextricably confounded with the changes in the number of animals within a group. For example, addition of an unfamiliar animal to a group is almost certain to precipitate intense aggression directed toward the intruder (unless it is an infant); withdrawal of a male or other dominant animal may also precipitate violence. Because the longitudinal changes are chiefly *qualitative* in nature, no one would seriously regard such a study using primates as providing evidence for crowding influences on aggressive behavior.

Cross-sectional studies. The obvious confounds of longitudinal studies of social-density effects, and the difficulty of finding many groups of monkeys housed under similar conditions for use in cross-sectional studies, seriously delayed analysis of the question of optimal group size for captive primates. Finally, we (Erwin and Erwin, 1976) were able to do a study using large numbers of groups of pigtail macaques housed under identical spatial conditions. The results of that study are reported later in this chapter. A note of caution is in order regarding the generalizability of studies in which the amount and quality of space is constant, i.e., *the optimal group size under one spatial or other qualitative condition is not necessarily the optimal size for another condition.*

The Concept of Interaction Potential

As mentioned at the beginning of the preceding section, it is necessary to generate some predictions about the relative influences of spatial and social changes as they affect density and especially as they influence behavior. Thus, it is the *translation* of social and spatial variation into observable aggressive behavior that concerns me here. While the simple models already presented suggest a direct linear relationship between both spatial variation and social variation as they affect density and behavior, it would be implausible to suggest, returning to the example of the baboons in the 400-m² corral, that the spatially mediated and socially mediated influences that resulted in identical changes in density would result in comparable changes in incidence of aggressive behavior. Most of us who have worked with primates would predict that 100 baboons fight more than 10 baboons under identical conditions of density. Of course, they would fight more–there are more of them to do so! Then, should they fight the same amount per capita so that we would expect the same percentage of subjects to sustain injuries? Even that would seem unlikely to us. There is a sound theoretical basis for the belief that crowding based on increases in the number of individuals present in a specific area is more influential than crowding based on spatial fac-

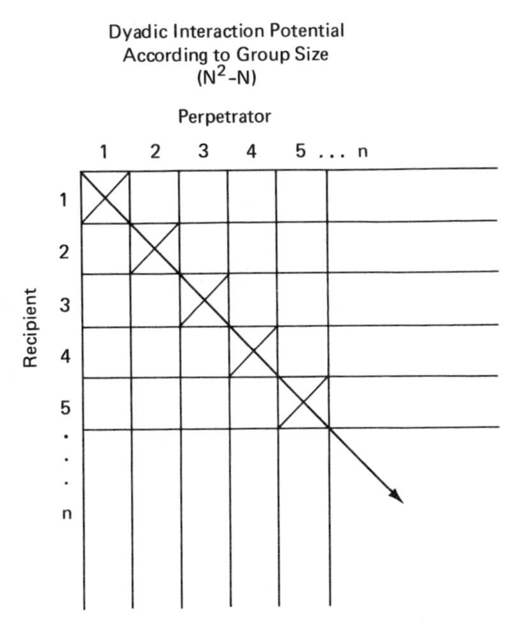

Fig. 6-2. Matrix illustrating derivation of dyadic-interaction potential for groups of differing sizes.

tors alone. I propose to demonstrate the utility of a concept I call *interaction potential* in assisting in the construction of expected values to interface behavioral data with density considerations. The concept of interaction potential is straightforward and certainly not new. It is simply based on the principle that moving bodies in confined space have a greater likelihood of interacting as the space is decreased or the number of objects in the space increases. The likelihood of interaction is also increased as a function of activity rates of the moving objects.

Is interaction potential affected differently by spatial variation than by variation in the number of individuals in a group? Yes. Should we reduce the amount of space available to a group by one-half, we would double the density and the interaction potential would also be doubled; that is, the likelihood of an interaction (of whatever kind) would increase in linear fashion. The picture is different, however, for changes in interaction potential related to variability in the number of group members.

Consider that each member of a group can be involved in a dyadic interaction in two ways, as the initiator or *perpetrator* of an act toward another individual, and also as the *recipient* of such an act. The matrix in Fig. 6-2 shows the rationale for dyadic-interaction potential as a function of group size. Thus,

$$\mathrm{IP}_d(fN) = N^2 - N$$

and, consequently, the expected value for dyadic interactions among group members is a positively accelerated function of group size (see Fig. 6-3).

The utility of employing relative dyadic-interaction potential is primarily that of comparing groups of different sizes in order to decide whether or not differences in frequency of behavior between groups are primarily based on *opportunity* to interact or some other factor. The matter of activity (rapidity of motion) among objects may be added as another dimension of interaction potential; the more active group members are, the more likely they are to interact. Thus, motivational forces that are activating often tend to increase the likelihood of interactions in general. Thus, *arousal* contributes to aggression in groups. The concept of interaction potential was derived from data reported later in this chapter.

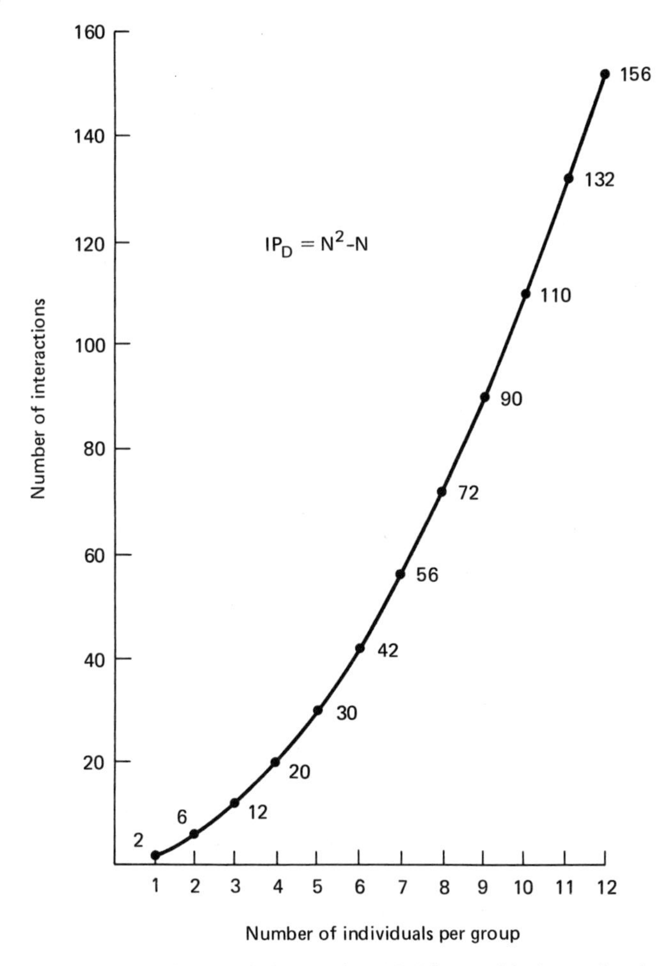

Fig. 6-3. Relative potential for dyadic interaction (IP_d) is a positively accelerating function of group size.

Critical Issues in Crowding Research

Population Control. Among the most important issues in research on the effects of density on behavior are those related to population control mechanisms. The crowding research of Calhoun (1961) and Christian (1955) has emphasized the role of crowding in natural population controls. When rodent populations become very dense, reproduction disturbances arise, including cessation of maternal care, rejection, cannabalism, and embryonic resorbtion. There are often clear physiologic indications of heightened stress, including increased adrenal weights and increased levels of corticosteroids in the blood. These changes are primarily associated with longitudinal studies involving unidirectional increases in density due to birth into groups residing in confined spaces. Aggressive behavior typically increases among rodents in this kind of situation. Essentially no evidence is available for nonhuman primates with regard to the role of longitudinal population density increases in relation to aggressive behavior, stress, or population control. Eaton et al. (1977), however, have reported that intragroup aggression among Japanese macaques at the Oregon Regional Primate Research Center has not increased relative to the population, which doubled in size over a four-year period of behavioral study. Density in that group is, however, very low by comparison with many other captive groups of nonhuman primates. As mentioned earlier, in the Elton study (Chapter 5, this volume) reproductive activity continued even under the most crowded conditions, but aggressive behavior directed toward juveniles became relatively common in the highest density situation. Although there may be some natural population controls for primates (e.g., delayed puberty in malnourished individuals), there is no convincing evidence that any population control mechanism is mediated by aggressive behavior, or any other behavior. The issue is by no means closed, however, because few studies have been attempted.

Short-Term Spatial Change. Studies of short-term changes in amount of space available have been limited in number and quality. Southwick (1967) reduced the amount of space available to a group of rhesus monkeys in captivity to one-half that previously available. He found no significant increase in aggressive behavior among group members. The results of the Southwick (1967) study have been frequently quoted as supporting the hypothesis that crowding increases aggression. That assertion is untrue. The study involved placing a barrier across the middle of an enclosure. At first, the gate in the barrier was left open; then, the animals were crowded into one-half of the enclosure and the gate was closed. While the frequency of aggression was twice as high under the crowded condition as under the *baseline* condition prior to installation of the barrier, there was no significant difference between the crowded condition and the phase just prior to it (when the barrier was in place but animals had access to both sides). In fact, there was a *larger* increase over baseline when the barrier was in place with the gate open than between the gate-open phase and the crowded phase following it. There were also qualitative, as well as quantitative changes in the environment assessed by Southwick, and the reversal design necessary to evaluate spatial differences was not employed. Despite the methodological weaknesses of the crowding aspect of Southwick's study, it *did* address the problem of crowding effects on behavior at a time when other investigators of primate behavior could offer only speculation regarding these effects, and Southwick should be complimented for his effort. While a dyadic-interaction potential model would have predicted exactly the result obtained by Southwick with regard to the baseline (no barrier) condition, as compared with the crowded condition, it must be conceded that placement of the barrier apparently accounted for most of the change in aggressive behavior that occurred in that experiment.

Alexander and Roth (1971) studied the effects of repeated crowding of Japanese monkeys from a large corral into a small enclosure at the Oregon Regional Primate Research Center. They found that aggression among males increased under the crowded condition, but that female-female aggressive interactions were significantly reduced in the small enclosure relative to the large corral. While there were many qualitative differences between the two environments, such as trees and concrete pipes in the corral and cement and wire in the pen, all qualitative differences would have been predicted to act in favor of lower aggression in the large enclosure; that result was true only for adult male-male interactions.

Studies of short-term effects of spatial change have not been definitive, and much more research on this topic is necessary before a complete description will be possible.

Optimal Group Size. As the number of primates in captivity for domestic breeding and display continues to rise, and as the need to protect wild troops and habitats increases, it is essential that decisions be made regarding optimal group size for specific areas and conditions. Yet, very little information is now available on this topic. Some of my own research on this topic is reported below.

Effects of Short-Term Spatial Crowding on Aggression

As mentioned earlier, the research reported here on the effects of crowding on aggression at the Regional Primate Research Center Field Station, Medical Lake, Washington, was initiated in response to a matter of primate husbandry. A large number of pigtail monkeys (*Macaca nemestrina*) were housed in groups in a building that was designed and used formerly as a prison. The cells were renovated to provide gang caging for the monkeys. At the time this project was initiated, each group ordinarily had access to two adjacent cells (two-room suites) via a small shuttle door. Every day, each group was crowded first into one room and then into the other while the empty room of the suite was cleaned. I was concerned that this daily crowding might have been producing heightened stress or aggression with consequent increases in disease or injury. These projects were reported by Anderson et al. (1977).

Experiment 1. The subjects for the first experiment were ninety-eight pigtail monkeys housed in six groups. Each room of each suite was 2.2 X 3.1 X 2.8 m (14 m^2 of floor space for both rooms and about 7 m^2 each).

Absolute frequencies of ten classes of agonistic behavior (hit, grab, push, bite, chase, open-mouth threat, bark, grimace, screech, and crouch) were scored by experienced primate behavior observers. The scoring procedure allowed identification of the perpetrator and recipient of each act as an adult male, adult female, or infant. Agonistic behaviors were monitored under two conditions, the normal two-room situation and the one-room (crowded) situation. Two sequences of observation sessions were employed: the one-two-one sequence in which behavior was measured first in one room of the suite, then in both rooms, and then again in one room; and the two-one-two sequence in which observations were made first in two rooms, then in one, and finally in the two-room situation. Each segment of each sequence lasted for twenty minutes and in all cases the two sets of observations were done on different (not necessarily consecutive) days. Interobserver reliability for all behaviors exceeded 0.90.

Results. Interactions involving males or infants were infrequent relative to those among adult females, and all the significant findings of the experiment pertained to female-female relationships. Interactions among females were affected in an astonishing manner. *Con-*

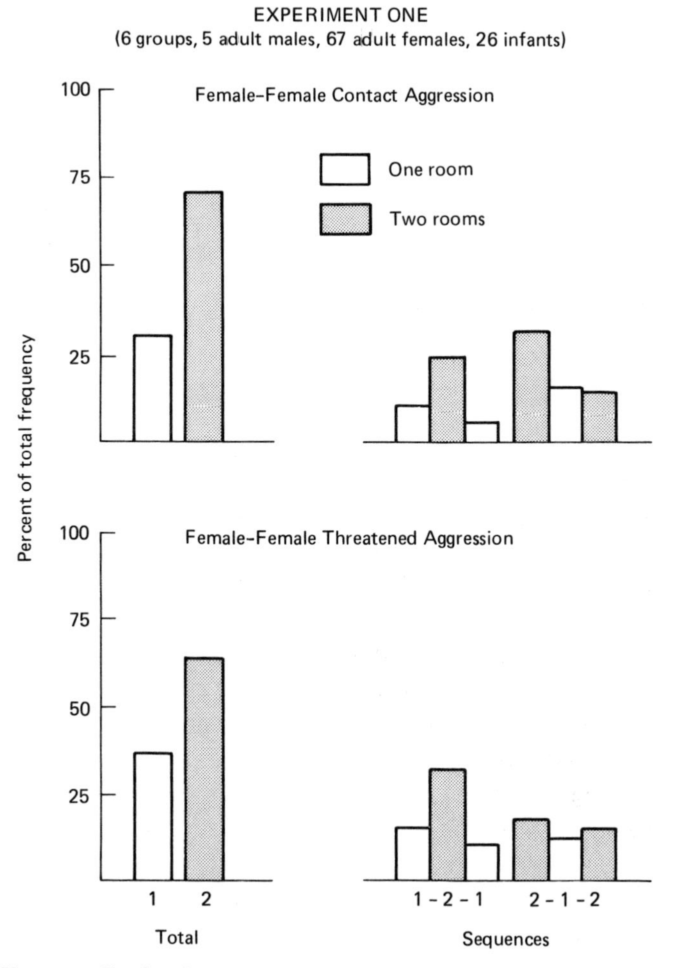

Fig. 6-4. Unexpectedly, females were more aggressive when they had access to two rooms than when they were crowded into one.

tact aggression (grab, push, hit, and bite) and *threatened aggression* (open-mouth threat) occurred *less* frequently in the crowded condition than in the relatively less crowded one (see Fig. 6-4).

Experiment 2. We were so surprised by the outcome of Experiment 1 that we felt compelled to redo the study using more groups and additional controls. Because there were some inconsistencies between the results of the one-two-one and the two-one-two sequences, we suspected that *spatial change per se* might have influenced the results independently of (or in interaction with) spatial quantity. We responded by adding two conditions that did not involve any spatial change. The no-change sequences were, of course, one-one-one and two-two-two, with twenty minutes per segment for a total of one hour per sequence. Of course, the one-two-one and two-one-two segments were retained. Observations were done with one spatial-change sequence and one no-change sequence each day for two consecutive days, balanced across type of sequence and days. The subjects were 109 pigtail macaques in 8 groups, none of whom had been subjects in the previous study. All groups in this study included one male and several females (including unweaned progeny).

Results. The results of Experiment 2 were consistent and convincing, and again opposite initial expectations. Contact aggression and threatened aggression among adult females were more than three times as frequent in the two-room as in the one-room situation. As before, there were no significant differences in interactions involving adult males or infants.

Contact aggression among females was significantly more frequent in two rooms than in one room under the spatial-change condition (Wilcoxon $T = 0.5$; $P < 0.02$) (see Fig. 6-5) and under the no-change condition (Wilcoxon $T = 2$; $P < 0.02$) (see Fig. 6-6). The frequency of contact aggression did not differ across spatial-change conditions for one room or two rooms. A comparison across spatial-change conditions revealed that the one-room no-change results differed from the two-room spatial-change results (Wilcoxon $T = 0.5$; $P < 0.02$), and the one-room spatial-change results differed from the two-room no-change results (Wilcoxon $T = 0$; $P < 0.01$).

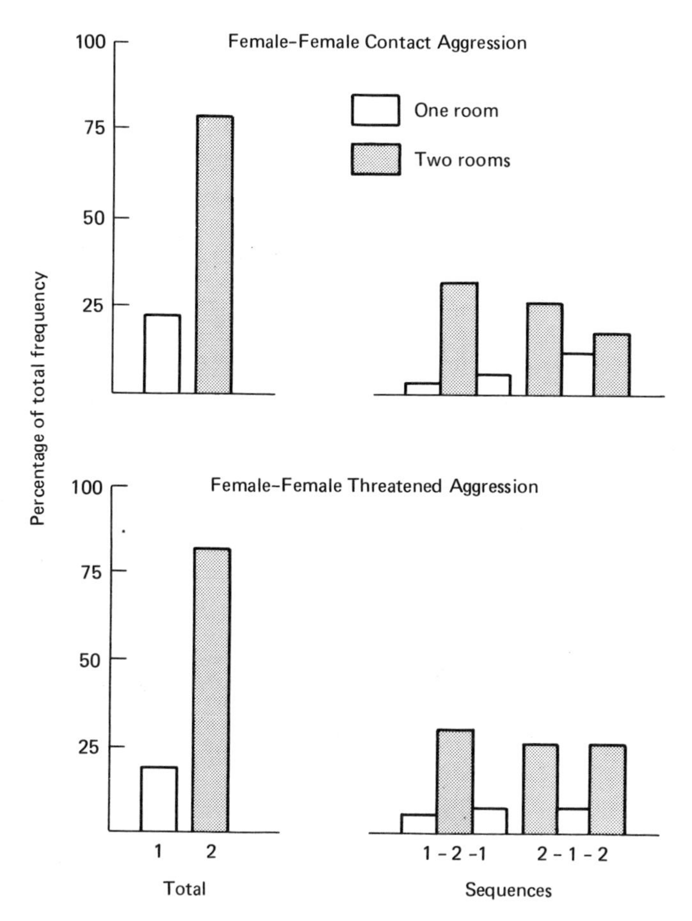

Fig. 6-5. The result shown in Fig. 6-4 was replicated using more subjects and a refined experimental design.

Threatened aggression among females was significantly more frequent in two rooms than in one room under both conditions (Wilcoxon $T = 0$; $P < 0.01$). The frequency of threatened aggression did not differ across spatial-change conditions for one room or two rooms. More threatened aggression occurred in the one-room no-change condition than in the two-room change condition (Wilcoxon $T = 0; P <$

0.01) and threatened aggression in the one-room spatial-change condition significantly exceeded that during the two-room no-change condition (see Figs. 6-5 and 6-6).

Thus, it was clear that the risk of violence among females was substantially higher when groups had access to both rooms than when they were crowded into one room. Crowding did not have the same effect as in the Southwick (1967) study of one group of rhesus mon-

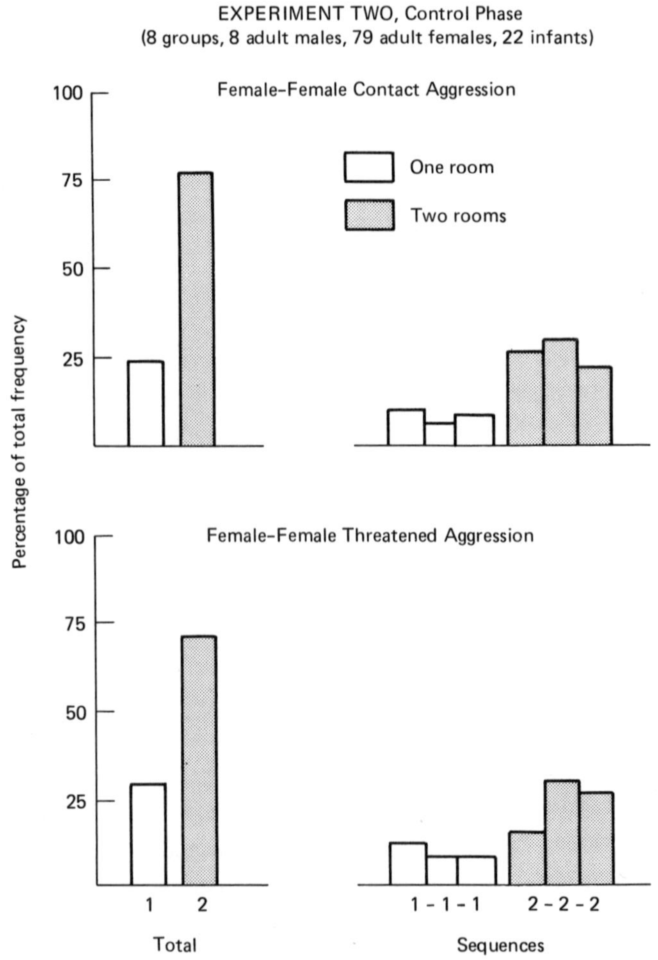

Fig. 6-6. A control phase of the replication showed that the result did not depend on spatial change.

Plate I

Some characteristic facial expressions of rhesus monkeys. (a) Threat. (b) Grimace.

Plate II

Some characteristic expressions of rhesus monkeys. (c) Yawn. (d) The expression associated with clear-call vocalization. (Photos: W. K. Redican.)

Plate III

An adult male and infant rhesus dyad. (a) Typical proximity established and maintained by the infant. (b) Female infant watches adult male yawn (note the potentially destructive canine teeth). (c) An infant rides "jockey style" on an adult male's back. (d) A seldom seen ventral-ventral contact posture, with support of the infant by the adult male care giver. (Photos: W. K. Redican.)

Plate IV

a. Ventral-ventral posture typical of rhesus mother-infant dyad. (Photo: W. K. Redican.)

b. Adult males together. The wild-born male retreats from the lab-born (isolate-reared) male in apparent confusion and submission. (Photo: N. Fittinghoff.)

c. Interspecific copulation between Romeo Rhesus and Julie Baboon. The resulting offspring was to be called a "MacBoon." (Photo: T. Maple.)

d. Juvenile rhesus and baboons formed affectional bonds. (Photo: T. Maple.)

Plate V

a. Adult male attempts to remove barrier after separation from infant. (Photo: W. K. Redican.)
b. Young adult female rhesus grooms male agemate during reunion following two-year separation. Affectional ties withstood the lengthy separation despite substantial maturational changes in appearance. (Photo: J. Erwin.)
c. Rhesus and baboon juveniles attempted to remove the barrier during the separation phase of Maple's interspecies attachment study. (Photo: J. Erwin.)

Plate VI

Male rhesus monkeys spent one year together as juveniles; nearly three years later they preferred each other's company to that of a highly receptive, but unfamiliar female. (a) The males mounted each other with anal penetration, and (b) attacked and injured the female. (Photos: T. Maple.)

Plate VII

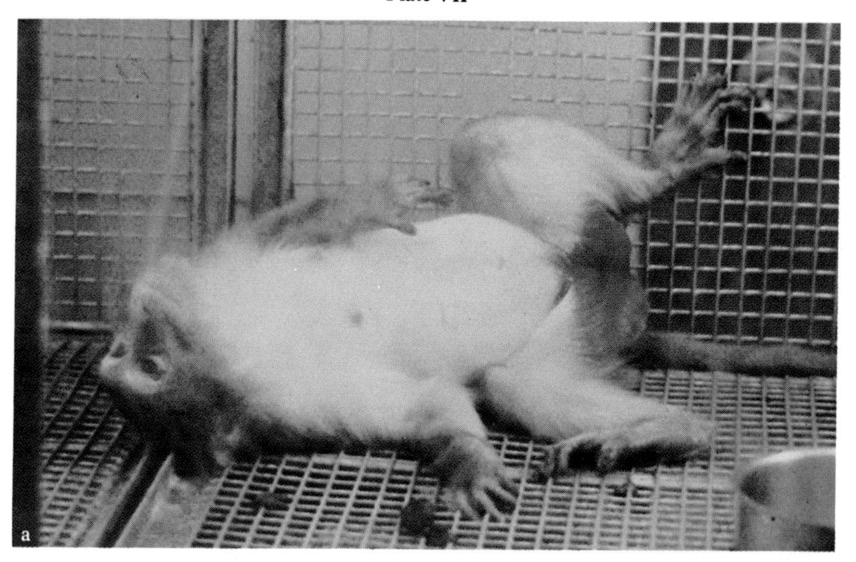

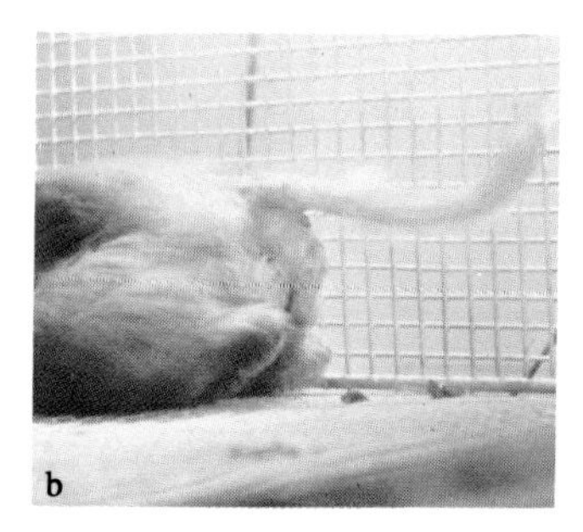

Rhesus female in labor. (a) Resting phase. (b) Facial presentation of infant. (Photos: W. K. Redican.)

c. Adult female grooms other female during labor. (Photo: W. K. Redican.)

a. Adult male watches female immediately following delivery of infant; female protects infant from male. (Photo: W. K. Redican.)

b. Adult female licks amniotic fluid from her hands and the neonate prior to ingesting the placenta. (Photo: W. K. Redican.)

Fig. 7-1. Adult Saimiri.

keys. Crowding did not increase aggression, but actually was related to *decreased* aggression among adult females. Subsequent studies described below revealed the basis of these unexpected results.

Crowding Based on Social Density

In the studies of spatial variation and aggressive behavior reported above, the number of females per group varied between six and fifteen. Thus, it was possible to test for a possible correlation between the number of potential interactors and the frequency of various kinds of interactions. The frequency of contact aggression per capita among females was significantly correlated with the number of females per group (Spearman's $rho = 0.62$; $P < 0.02$) (based on data for all one-two-one and two-one-two sequences from the fourteen groups involved in Experiments 1 and 2). Only the relationship between *contact* aggression and group size was significant; threatened aggression was not correlated with group size. While the relationship between the size of the group and the frequency of contact aggression was expected, the slightly negative correlation between threatened aggression and group size was puzzling. I was not entirely confident of the results regarding social density due to the correlative nature of the data and the fact that there were data for only fourteen groups of monkeys. The following large-scale survey was conducted to provide additional information on social-density influences on aggression (Erwin and Erwin, 1976).

Method. The subjects for the survey were all available group-housed pigtail monkeys at the RPRC Field Station. Each available group was surveyed four times over a four-month period. The scoring procedure was identical to that used in the spatial-density experiments reported above. Each observation period lasted for twenty minutes. Because 85 percent of the groups changed in number of occupants from one observation to the next, each observation was considered separately. Observation sessions were then classified according to the number of adult females residing in them at the time of the observation. In order to guarantee acceptable levels of reliability within group size, data were used *only* for group sizes for which six or more observations were available. A total of seventy-two group observations for group sizes ranging from eight to thirteen adult females met the criterion. The results of the study are illustrated in Fig. 6-7.

As in the previous study, contact aggression was highly correlated with group size. The mean frequency of contact aggression among females was directly related to group size (Spearman's *rho* = 0.99). At least one group of each size displayed *no* contact aggression, but the percentage of groups of each size that displayed some contact aggression was also highly correlated with the *per capita* frequency of contact aggression (Spearman's *rho* = 0.99). Even when scores were adjusted for *interaction potential* (as shown in Fig. 6-7), the positive correlation between group size and frequency of contact aggression remained strong. Thus, the putative relationship between contact aggression and group size was even stronger than would have been predicted on the basis of increased opportunity for interaction, and

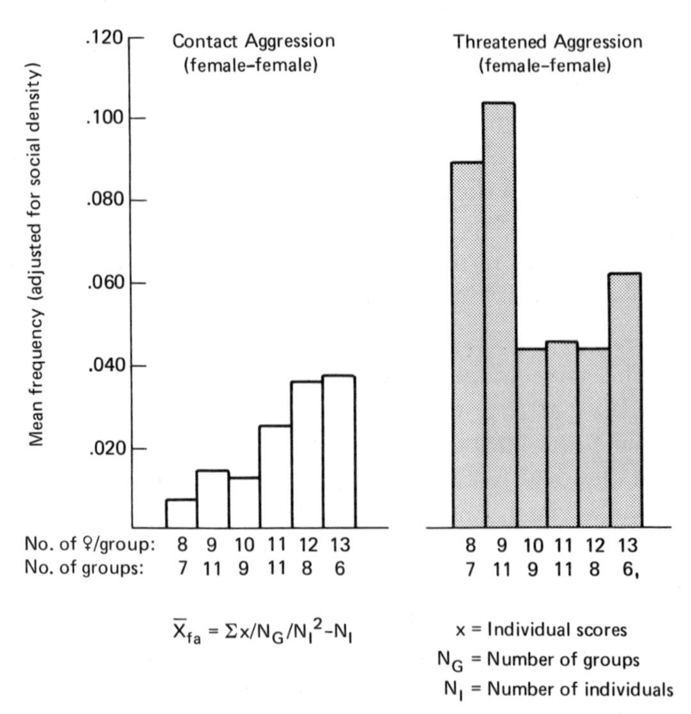

Fig. 6-7. The amount of contact aggression among adult females was highly correlated with the number of females per group, even after the data were adjusted for interaction potential.

probably reflected additional contact aggression generated by social stress.

On the other hand, threatened aggression did not exhibit the same strong relationship with social density as was seen in contact aggression. In fact, there were high levels of threatened aggression only at the lower densities. Because threatened aggression ordinarily served a communicative function, probably to reduce overt physical contact aggression, it appears that the reduction in threatened aggression signaled a communicative breakdown—that contact aggression continued to increase (even more rapidly than did density) when threatened aggression ceased to function appropriately.

Summary

The studies described here offer a firm basis for believing that group size is an especially important determinant of aggressive behavior in macaques. The expected rates of interaction generated by an interaction potential model are surpassed by observed values of contact aggression. The data suggest that nine or ten females per harem group is a maximum size for the captive conditions under which the study was done. On the other hand, the studies on spatial variation suggested that the more crowded condition actually lowered risk of trauma due to contact aggression. In fact, analysis of colony records of bite-wound treatments revealed that fewer treatments had been required prior to adoption of the two-room-suite housing policy (while groups were housed in single groups). The one-room-housing strategy was readopted and the trauma rate was considerably reduced. The reduction in trauma could not be entirely accounted for by the change in housing; a number of other factors were changed at about the same time based on other research findings and clinical judgments.

CONTROL ROLES AND AGGRESSION

It seemed probable that some aspects of group composition other than the number of females per group might have influenced aggression as well. Some surveys indicated that female aggressive interactions were less frequent in groups containing males than in those containing no males. We were considering housing some additional females in all-female groups for contraceptive reasons or to protect newborns from assault by adult males.

In the first survey (Sackett et al., 1975) we used six groups of *M. nemestrina* (containing sixty-seven individuals). Three of the groups contained adult males; three did not. Eight females had infants with them. Each group was observed twice with a different observer for each observation. Each observer was blind to the results of the other observers. On both the original and the replicate observations, there

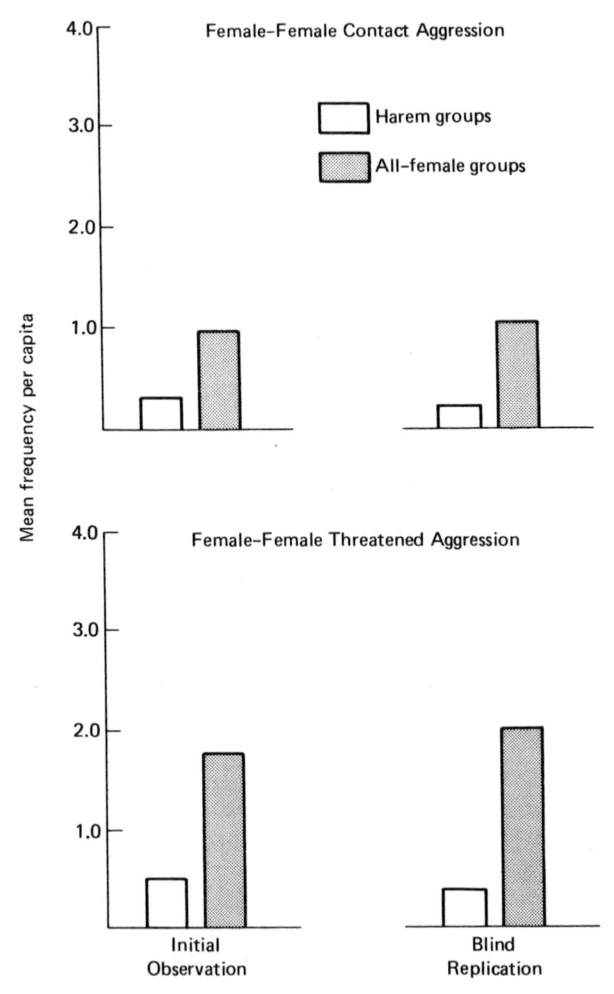

Fig. 6-8. Females in all-female groups were more aggressive among themselves than were females in groups containing adult males.

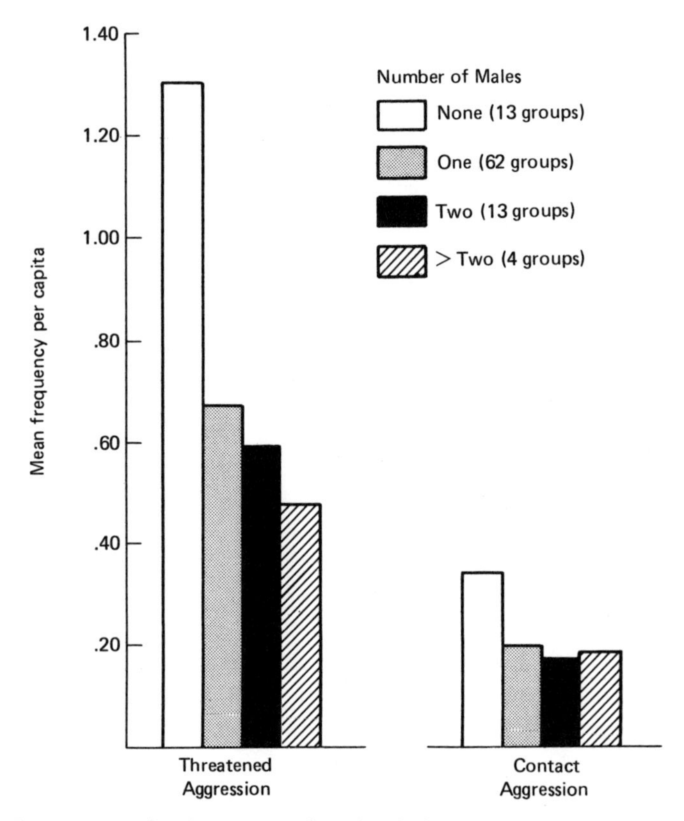

Fig. 6-9. The presence of at least one adult male within a group was associated with a considerable reduction in incidence of aggressive interactions among females relative to groups containing only adult females and their offspring.

was considerably more aggression among females in groups containing no male (see Fig. 6-8) than in groups containing one male.

Another survey was conducted, this time involving 20 groups containing 203 pigtail macaques (15 adult males, 166 adult females, and 22 infants) (Dazey et al., 1977). Fifteen groups contained one adult male each, and five contained no adult male. Ten of the groups contained no infants, and the other ten contained one to three infants each. Again, observations were done with one blind replication. As in the previous survey, females exhibited significantly less contact or threatened aggression in groups containing a male than in all-female groups. Noncontact aggression was significantly less frequent in groups containing adult males if infants were also present than if no infants were present. Thus, the results of this survey substantiated those of the previous study and lent additional credence to the possibility that males inhibit aggression among females.

It seemed plausible that the presence of additional males in groups might further inhibit aggression among females. The data obtained in the large survey of aggression according to group size, which had previously been analyzed in terms of effects of number of females per group relative to frequencies of aggressive behaviors, was replotted in terms of female-female aggression according to the number of adult males per group. The results indicated (Fig. 6-9) that the presence of one male was sufficient to inhibit threatened or contact aggression considerably, and the presence of more than one adult male had relatively little effect, particularly with regard to contact aggression.

To further substantiate this phenomenon, we (Oswald and Erwin, 1976) experimentally removed the resident male from each of six pigtail monkey groups (six males, fifty-one females, and seven infants). The experiment involved a twenty-minute pretest (male present), a twenty-minute test (male absent), and a twenty-minute posttest (male again present). Removal of adult males from groups resulted in dramatic increases in contact aggression among adult females. Reintroduction reduced female-female aggression to preseparation levels (see Fig. 6-10).

In every survey or experiment of this series there was very clear evidence that the mere presence of males in groups inhibited aggression among adult females. Other investigators have reported active interference of males in agonistic encounters among other group members (Bernstein, 1964; Smith, 1973) and a previous study involving temporary removal of a dominant male demonstrated an effect similar to that reported here (Tokuda and Jensen, 1968). We have also seen active intervention in female-female encounters, but, far more frequently, a male merely glances at a female involved in an aggressive interaction, and the aggressive bout ends.

This effect is almost certainly the basis of the paradoxical effect of spatial change on aggression described earlier in this chapter. Although the two-room suites at the RPRC Field Station provided twice as much space as did the single-room situation, there was a subtle qualitative difference between the one-room and the two-room

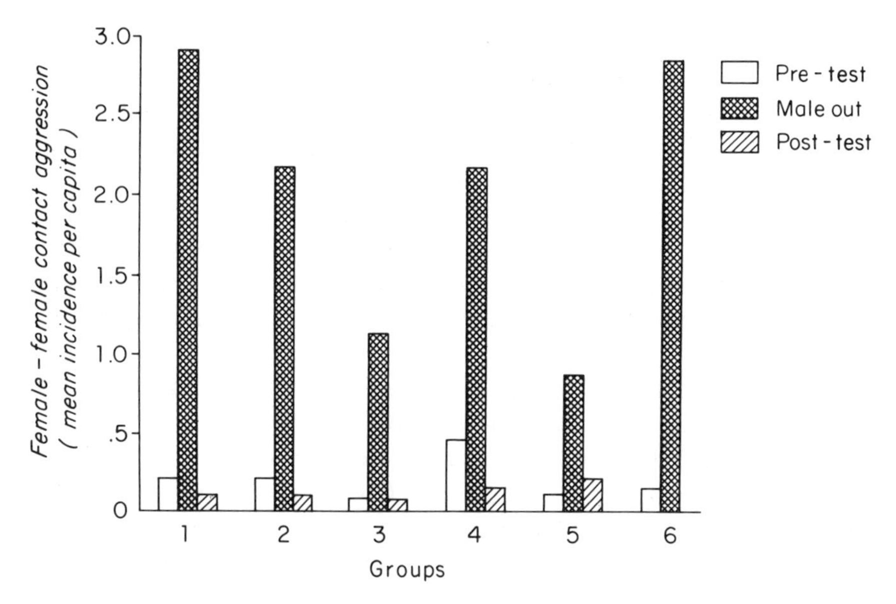

Fig. 6-10. Removal of males from groups resulted in increased violence among females; reinstatement of males in groups reduced violence among females to preseparation levels.

situation. If a group contained a male, he could not be in both rooms at once, thus, he could only inhibit aggression in one of the rooms. We discovered that aggressive encounters in the room away from the male were typically of longer duration and consisted of successive bites and grabs. Aggressive encounters among females near the male usually consisted of a single act, such as a grab or bite. There seldom was any reciprocal fighting by females in the direct presence of males. It appears, in fact, as if some adult females remained near the males specifically for protection from other females.

AGGRESSIVE BEHAVIOR AND "COVER"

Recognition of the interface between aggressive behavior and the quality of captive environments, such as the interface between the male control role and the use of additional space outside the realm of this influence, is essential to the design of effective captive environments. Wilson (1971) found that provision of visual barriers for use as "cover" by Japanese monkeys in captivity elicited escape responses that were sometimes effective, but that overall levels of aggressive behavior were not reduced in that situation relative to a barrier-free space. On the basis of our experience with two-room versus one-room housing, we attempted to assess the effects of a small escape space as cover. It was clear that the provision of extra space with too much visual isolation promoted rather than inhibited aggressive behavior. We reasoned that escapability and privacy (in the sense that the escape space should not be able to accommodate fighting animals) should be the main requirements for structures to provide cover.

We (Erwin et al., 1976) chose to use concrete cylinders to provide cover for pigtail macaques at the RPRC Field Station. The cylinders were about 0.5 m in diameter and were open at both ends. Metal braces were fashioned to keep the cylinders against one wall of the room. The subjects for this study were ninety-eight monkeys, seven adult males, sixty-nine adult females, and twenty-two infants, housed in six social groups. Most of the time the animals had access to both rooms of their two-room suites, but for a short time each day the animals were crowded into one-half their normal space while the unoccupied room of their suite was cleaned.

The study was conducted in two phases, a pretest or baseline phase, and an experimental or test phase. During the pretest phase of the study, animals were observed with no change in their physical environments; at the beginning of the test phase, cylinders were introduced into the rooms.

The pretest and test phases of the study were each five days long. Each group was observed for twenty minutes each day by a pair of experienced observers (one member of each pair monitored behavior occurring in each of the two-room suites). Data were recorded in the same fashion as for the other studies of this series. Four of the groups remained reasonably stable throughout the period of the study, but two underwent mild and major changes, respectively.

In those groups that remained stable, aggressive interactions were considerably reduced during the presence of cover. The disruption of the other groups resulted in increased violence (Fig. 6-11). In one case, referred to earlier, violence was severe and devastating (see Fig. 6-1). While the availability of cover of the right type can apparently reduce violence, it clearly was unable to overcome the aggressive motivation associated with unstable newly formed groups.

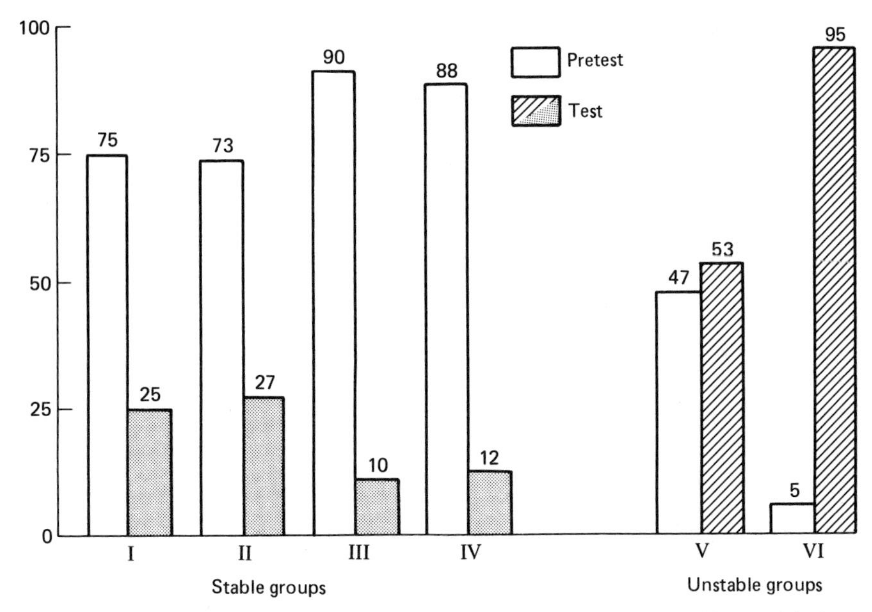

Fig. 6-11. Provision of cover reduced violence in groups that remained stable but failed to overshadow the effects of group disruption.

VIOLENCE AND GROUP FORMATION

Formation of new groups has resulted in more casualties than any other procedure at the RPRC Field Station. In the past, it has been difficult to avoid occasional reorganization and reconstitution of groups, because subjects for experimental use have been withdrawn and later reintroduced. Animals withdrawn for treatment or disease have also returned to groups as necessary. Since group formation was found to place animals at very high risk for trauma, it was especially urgent that we evaluate group formation strategies in order to find better procedures that would reduce risk.

Earlier tests by other investigators (Bernstein, 1964; Bernstein and Gordon, 1974) had found that sequential introduction of individuals into groups resulted in more aggression and injuries than did simultaneous release of all animals. When reorganization of groups was required about three years ago, we systematically compared three simultaneous introduction techniques (see Erwin, 1976; Erwin, 1977a).

Nine groups were formed in two-room suites using 9 adult males (one per group), 121 adult females, and 21 infants. Groups were formed according to three general strategies: 1) merger of females from two existing groups; 2) merger of subgroups of three or four familiar females each; and 3) introduction of females that were familiar with no more than one other member of the new group (for the most part these females were not familiar with any other individual in the group). In one group of each of the first two types, males were introduced who were familiar with members of one group of subgroup. For all other groups males were unfamiliar, as far as we could discern, with any of the other females.

The merger of subgroups resulted in more casualties than did either of the other strategies (eleven females or infants sustained bite wounds in groups formed from subgroups within the first month after formation). No injuries occurred in groups formed of total strangers (although there was much fighting among these groups, the fighting was apparently not very serious from the standpoint of bite wounds). Two of the groups formed by merger of existing groups had some animals injured in them (though not as many as in the subgroup mergers), and one merger of existing groups resulted in no injuries. *No female or infant familiar to a male was injured within the first month after group formation.* Animals familiar with one another formed coalitions against others. This resulted in sustained fighting and many injuries except to females who joined in coalitions with (familiar) males. In *M. nemestrina* and *M. fascicularis*, infant trauma due to addition of an unfamiliar male to a group is a major cause of infant mortality (Dazey and Erwin, 1976; Erwin, 1977b).

CONCLUSIONS

Many of the results I have reported here were unexpected—in several cases they were diametrically opposed to obvious prediction. I believe that is very encouraging. By careful, quantified observation, it is possible to find out about things as they are. While things may not exist everywhere in the same state, they can be found to be consistent within specific circumstances, and the knowledge derived can be applied directly in that setting and generalized to and tested in other settings. It is premature to say that the findings reported in this chapter will extend to all circumstances, but I will offer the following statements that may lead to verification and generality.

Social-emotional bonds may develop among nonhuman primates of any age or either sex, within, as well as between species; this must be kept in mind when housing animals alone or together at every stage of life, or when contemplating the separation from or addition of animals to groups. Remember that social bonds may be exclusive and that animals may not accept strangers they would ordinarily tolerate if they are in the presence of another animal with which they have had previous experience.

Social disruption is probably the most devastating problem encountered by domestic breeders and displayers of primates. Every effort should be made to establish breeding colonies that allow maintenance of long-term bonds. Artificial additions to groups should be avoided; removal can also be disastrous, particularly the removal of a dominant male. Harvesting should, wherever possible, reflect the typical mortality patterns; that is, a large proportion of juvenile males can probably be removed safely and without disruption because they would ordinarily be peripheralized or killed anyway (Drickamer, 1974). In display settings, such animals should also be removed as they would ordinarily be excluded from the core group.

Spatial factors can be very important in the captive maintenance of primates. As in the case of the one-room versus two-room environments, common sense predictions often are not accurate. There is no substitute for actual testing of environments to see how the animals will respond. Observation with quantification adds power to decision making. Evidence is the strongest justification for implementing change or maintaining the status quo. The point of the one-room versus two-room crowding study is not that crowded animals are sometimes better off; the important point is that husbandry decisions should be scientifically based. There is currently a great deal of pressure to specify minimum spatial requirements for primates. Unfortunately, there is almost no evidence upon which to base the decision. The data reported here have shown that social factors usually outweigh spatial factors. This does not imply, as my good friend Tim Beamer suggested, that the ideal state of affairs for primate social housing would be the inclusion of large groups in small pickle jars to avoid the detrimental effects of aggressive behavior.

REFERENCES

Alexander, B. and Roth, E. The effects of acute crowding on aggressive behavior of Japanese monkeys. *Behaviour* **39**: 73-88 (1971).

Anderson, B., Erwin, N., Flynn, D., Lewis, L., and Erwin, J. Effects of short-term crowding on aggression in captive groups of pigtail monkeys (*Macaca nemestrina*). *Aggressive Behav.* **3**: 33–46 (1977).

Ardrey, R. *The Territorial Imperative.* New York: Atheneum, 1966.

Bernstein, I. The integration of rhesus monkeys introduced to a group. *Folia Primatol.* 2: 50–63 (1964).

Bernstein, I. and Gordon, T. The function of aggression in primate societies. *Am. Sci.* **62**: 304–311 (1974).

Calhoun, J. Phenomena associated with population density. *Proc. Nat. Acad. Sci.* **47**: 428–449 (1961).

Christian, J. Effect of population size on the adrenal glands and reproductive organs of male white mice. *Am. J. Physiol.* **181**: 477–480 (1955).

Dazey, J. and Erwin, J. Infant mortality in *Macaca nemestrina*: neonatal and postneonatal mortality at the Regional Primate Research Center Field Station at University of Washington, 1967–1974. *Theriogenology* **5**: 267–279 (1976).

Dazey, J., Kuyk, K., Oswald, M., Martenson, J., and Erwin, J. Effects of group composition on agonistic behavior of captive pigtail macaques, *Macaca nemestrina. Am. J. Phys. Anthropol.* **46**: 73–76 (1977).

Drickamer, L. A ten year summary of reproductive data for free-ranging *Macaca mulatta. Folia Primatol.* **21**: 61–30 (1974).

Eaton, G., Modahl, K., and Johnson, D. Aggressive behavior in a confined troop of Japanese macaques. Paper presented at the Annual Meeting of the Western Psychological Association, Seattle, Wash., 1977.

Erwin, J. Aggressive behavior of captive pigtail macaques: spatial conditions and social controls. *Lab. Primate Newsletter* **15**: 1–10 (1976).

Erwin, J. Factors influencing aggressive behavior and risk of trauma in the pigtail macaque (*Macaca nemestrina*). *Lab. Anim. Sci.* **27**: 541–547 (1977a).

Erwin, J. Infant mortality in *Macaca fascicularis*: neonatal and postneonatal mortality at the Regional Primate Research Center Field Station, University of Washington, 1967–1976. *Theriogenology* **7**: 357–366 (1977b).

Erwin, J. and Flett, M. Responses of rhesus monkeys to reunion after long-term separation: cross-sex pairings. *Psychol. Rep.* **35**: 171–174 (1974).

Erwin, J. and Mitchell, G. Initial heterosexual experiences of adolescent rhesus monkeys. *Arch. Sex. Behav.* **4**: 97–104 (1975).

Erwin, J., Brandt, E., and Mitchell, G. Attachment formation and separation in heterosexually naive preadult rhesus monkeys (*Macaca mulatta*). *Dev. Psychobiol.* **6**: 531–538 (1973).

Erwin, J., Mobaldi, J., and Mitchell, G. Separation of rhesus monkey juveniles of the same sex. *J. Abnorm. Psychol.* **78**: 134–139 (1971).

Erwin, J., Maple, T., Willott, J., and Mitchell, G. Persistent peer attachments of rhesus monkeys: responses to reunion after two years of separation. *Psychol. Rep.* **34**: 1179–1183 (1974).

Erwin, J., Anderson, B., Erwin, N., Lewis, L., and Flynn, D. Aggression in captive groups of pigtail monkeys: effects of provision of cover. *Percept. Mot. Skills* **42**: 319–324 (1976).

Erwin, N. and Erwin, J. Social density and aggression in captive groups of pigtail monkeys (*Macaca nemestrina*). *Appl. Anim. Ethol.* 2: 265–269 (1976).

Fairbanks, L., McGuire, M., and Kerber, W. Sex and aggression during rhesus monkey group formation. *Aggressive Behav.* **3**: 241–249 (1977).

Hall, K. R. L. Aggression in monkey and ape societies. In J. D. Carthy and F. J. Ebling (Eds.) *The Natural History of Aggression.* London: Institute of Biology, 1964, pp. 51–64.

Lorenz, K. *On Aggression.* New York: Harcourt, 1966.

Nagel, V. and Kummer, H. Variation in cercopithecoid aggressive behavior. In R. Holloway (Ed.) *Primate Aggression, Territoriality, and Xenophobia.* New York: Academic Press, 1974, pp. 159–184.

Oswald, M. and Erwin, J. Control of intragroup aggression by male pigtail monkeys (*Macaca nemestrina*). *Nature* **262**: 686–688 (1976).

Rowell, T. A quantitative comparison of the behaviour of a wild and a caged baboon group. *Anim. Behav.* **15**: 499–509 (1967).

Sackett, D., Oswald, M., and Erwin, J. Aggression among captive female pigtail monkeys in all-female and harem groups. *J. Biolog. Psychol.* **17**: 17–20 (1975).

Smith, E. O. A further description of the control role in pigtail macaques, *Macaca nemestrina. Primates* **14**: 413–419 (1973).

Southwick, C. An experimental study of intragroup agonistic behavior in rhesus monkeys (*Macaca mulatta*). *Behaviour* **28**: 182–209 (1967).

Southwick, C., Siddiqi, M., Farooqui, M., and Pal, B. Xenophobia among free-ranging rhesus groups in India. In R. Holloway (Ed.) *Primate Aggression, Territoriality, and Xenophobia.* New York: Academic Press, 1974. pp. 185–209.

Tokuda, K. and Jensen, G. The leader's role in controlling aggressive behavior in a monkey group. *Primates* **9**: 319–322 (1968).

Wilson, C. An experimental manipulation of the ecological variable "cover": Its influence on captive Japanese macaques (*Macaca fuscata*). Ph.D. dissertation, University of Washington, Seattle, 1971.

7
Titi and Squirrel Monkeys in a Novel Environment*

D. Munkenbeck Fragaszy

California Primate Research Center and Department of Psychology
University of California
Davis, Calif.

INTRODUCTION

The most general function of behavior is to allow effective adjustment to environmental change. Response to novelty is therefore an important dimension of the behavioral organization reflected in a species' general life-style (Fragaszy and Mason, 1978; Glickman and Sroges, 1966; Menzel, 1966, 1969). Species-typical modes of dealing with environmental novelty may contribute in significant ways to species differences in a variety of functional behavioral categories. For example, feeding and use of space are two classes of behavior which are likely to reflect differences in the way species assimilate and accommodate to environmental change.

Sympatric Species with Differing Life-Styles

Squirrel monkeys (*Saimiri sciureus*) (Fig. 7-1, Plate 8) and titi monkeys (*Callicebus moloch*) (Fig. 7-2) are sympatric in some regions of South America and apparently share, in many respects, dietary requirements and habitat resources (e.g., compare Mason, 1966, and Thorington, 1968). However, members of these species differ dis-

*The author wishes to thank Prof. William A. Mason for his continuing advice and support concerning this project; University of California, Davis, Social Science Data Service and Prof. N. Matloff for their aid in computerization of the data; and R. J. Fragaszy for his generous help in computer manipulations of the data. R. W. Summers and W. A. Mason provided helpful comments on the manuscript. Thanks also to J. Sano for lending photographic aid, and to V. Gonzalez and E. Faurot for their conscientious work on illustration maps. Funds for computer usage were provided by the University of California, Davis, Computer Center. This research was supported by an NSF predoctoral fellowship to the author, and by NIH/NSPHS grants HD06367 to William A. Mason, and RR00169 to the California Primate Research Center.

Fig. 7-2. An adult *Callicebus* male.

tinctly in several dimensions of behavior. Two of the most obvious contrasts are in social organization and in use of space. In the wild, *Saimiri* live in relatively large mixed-sex groups and range over an undefended area, whereas *Callicebus* live in monogamous family groups that remain within small defended territories. Contrasting social tendencies are maintained in a variety of laboratory environments. *Callicebus* typically form enduring attachments to a single opposite-sexed pairmate, while comparably housed *Saimiri* do not (Mason, 1971, 1974, Cubicciotti and Mason, 1975). Similarly, species differences in general activity level in the laboratory are consistent with travel patterns evident in the wild. *Saimiri* show a high level of activity, whereas *Callicebus* are relatively sedentary (Mason, 1966, 1974; Thorington, 1967).

Social Relationships. Considerable information is available for these species on relationships between individual adaptive behavior and social tendencies. In *Callicebus*, for example, presence and behavior of the pairmate were important factors in the expression of interest in a novel object in the home cage. Investigatory activities were coordinated between members of a pair and the level of such activities declined in the absence of the pairmate. In contrast, pairmates' activities were not coordinated in *Saimiri*, and absence of the pairmate had no depressing effect on investigatory activities (Fragaszy and Mason, 1978). Visual and spatial orientation to the pairmate was higher in *Callicebus*, as were distress responses to separation (Cubicciotti and Mason, 1975; Phillips and Mason, 1976). Feeding was a more social activity in *Callicebus* than in *Saimiri* (Fragaszy, 1978). Altogether, these findings suggest that social factors play a more varied and influential part in organizing the responses of *Callicebus* to the nonsocial environment than they do for *Saimiri.*

Nonsocial Influences. In addition, observation of these monkeys under a variety of circumstances suggests that salience of nonsocial features of the environment vary for the two species: *Saimiri* more frequently manipulated various objects in the environment than did *Callicebus* and were quicker to approach novel items; likewise, *Callicebus* performed less well than *Saimiri* on a visual learning task (Fragaszy, in preparation). *Callicebus* were more discriminating than *Saimiri*, however, in the food items they selected and consumed, although they accomplished these activities more slowly (Fragaszy, 1978).

Thus, available information indicates that the species attend to and utilize environmental resources, including information resources, differently. Behavioral contrasts reflecting different ordering of priorities were expected to be sharpened and amplified in a large, heterogeneous outdoor environment, a setting which created more stress and more behavioral options than most laboratory situations in which these species have been observed. The research was designed to provide descriptive data on initial familiarization/exploratory patterns and on temporal changes in these patterns in each species. Particular focus was placed on social orientation, as this was expected to play a different role in spatial-orientation behavior in *Callicebus* than in *Saimiri.*

AN EXPERIMENTAL STUDY OF USE OF SPACE

The research reported here contrasted responses of *Saimiri* and *Callicebus* monkeys to a novel, large, outdoor enclosure containing various food sources, travel surfaces, and visual structures. Information was sought on how each species integrates environmental information into its ongoing behavior, and on how species-typical behavioral strategies for dealing with strange environments contribute to the overall behavioral life-style of the species.

Methods

Subjects. Subjects were five adult male-female pairs each of *Saimiri* and *Callicebus.* All animals were wild caught and laboratory acclimated. Each male-female pair had been housed together for several years, except for one pair in each species. In *Callicebus*, the exceptional pair had been housed together for four months when first tested, and in *Saimiri*, for two weeks (each member of this pair went through adaptation with a familiar animal other than the one used in formal testing). One *Callicebus* pair was living with its two female offspring, which were left in the home cage during testing.

When not being tested, all pairs were housed in identical 1 m X 1 m X 3 m indoor cages containing food and water ad lib and equipped with wooden perches at four levels in the cage. These cages communicated with an outside area, approximately 4 m X 1 m X 2 m, with two 1 m X 0.5 m rectangular-perch systems running its length. Barriers outside and inside minimized visual access between adjacent pairs.

Apparatus. The test environment was an enclosure measuring approximately 88 m X 91 m X 10 m high (1 hectare), constructed of 2.5 cm poultry mesh supported by wooden poles (30 cm in diameter at the base) and beams. The poles supporting the roof and added landmarks were used as location aids to mark off 576 sections, 3.75 m on a side. The enclosure was planted with corn, sunflowers, and young trees (various species, 1–2 m in height) (see Fig. 7-3). The western edge of the enclosure was adjacent to an Old World monkey outdoor housing complex which was partially visible from the enclosure. The view in other directions was of flat, plowed land (two sides) or of a paved parking lot and buildings. The area in which monkeys of both species were housed was behind a row of buildings about 400 m to the northwest of the enclosure.

Apart from the trees and roof supports, the major structural feature inside the enclosure was a 49 m × 29 m grid of raised runways and six associated feeders (see Fig. 7-3). Sunflowers and tall vegetation were removed from the interior of the grid area. The grid was constructed of lengths of 2.5 cm × 10 cm (1 in. × 4 in.) lumber attached to 1 cm diameter vertical iron rods. The top of the runway was approximately 0.3 m above ground level. The runway was marked into 2 m sections with paint of different colors to provide location cues. Feeders were wood-frame boxes with solid roofs and floors and poultry-wire sides, measuring roughly 44 cm wide × 47 cm high × 37 cm deep. Each feeder was fitted with a sliding door (roughly 30 cm × 40 cm) operated remotely by means of a rope and pulley arrangement. With the door raised, the opening was approximately 30 cm × 44 cm. Two aluminum trays (15 cm × 10 cm × 5 cm) were attached to opposite walls inside the box. Prior to each trial, all trays were filled with three identical items from the following list: cherry tomatoes (quartered), dates, peanuts (in the shell), carrots, apples, bananas. The last three were cut in 1.3 cm cubes. Each feeder was baited with the same items throughout all trials. Feeders were adjacent to the runway and slightly above runway level (approximately 0.4 m).

Observers equipped with binoculars sat on a 1 m × 0.7m × 3 m platform near the center of the enclosure. A fifteen-second audible intervalometer was used in collecting location data, and an Esterline-Angus event recorder was used to score occurrence and duration of moving and feeding bouts. Male subjects were identified by a narrow (approximately 1 cm) colored leather belt.

Procedure. Before formal tests, each pair was given several practice sessions with a baited feeder in a familiar location. This continued

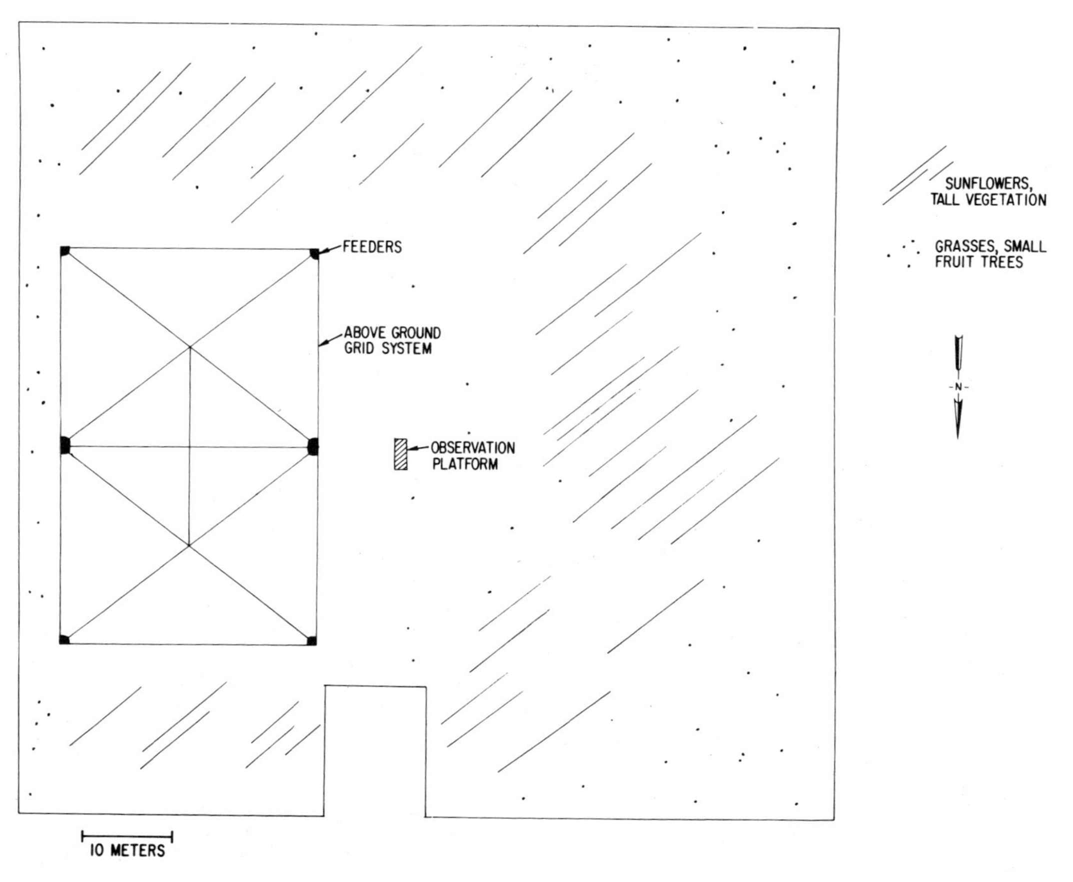

Fig. 7-3. Sketch of the release enclosure. The 49 m × 29-m aboveground grid and associated feeders are shown, as are major vegetational zones.

until the monkeys entered and fed freely, which usually required three to four sessions. One adaptation session was also given in the enclosure. The adaptation session in the enclosure involved placing both members of a pair together in a baited feeder and allowing them to remain there until they appeared calm and ate (usually this took one or two minutes), after which they were allowed to travel away from the feeder, then recaptured and placed in another feeder. This procedure was repeated with each of the six feeders. Each pair also had a one-hour "free" period in the enclosure before formal testing started.

Animals were tested in pairs, and were brought to the enclosure in a wire carrying cage. Testing occurred in the morning and early afternoon, before daily feeding, when the temperature was between 55°–85°F (13°–30°C) and the weather dry and calm. Before a trial, feeder trays were baited.

To start a trial, the pair was released from the carrying cage at the center of the grid system. Observations began immediately. One observer located both subjects at the onset of every fifteen-second interval, noted activity data, and indicated this information to a second person who served as recorder. Another observer operated the keyboard of an event recorder to time the onset and duration of travel and feeding bouts for both animals. At the end of the trial, notation was made of the amounts of food left in the feeders. Each pair was tested once per day at roughly one-week intervals for a total of five one-hour trials per pair. Trials occurred over an eight-week period in September, October, and November, 1975.

Behavioral Measures. The following data were collected for each subject at fifteen-second intervals: location (including in feeders), locomotion, following (one monkey moving in the same direction behind the other and within 15 m), social contact, and feeding. Feeding was differentiated into feeding on provisioned foods and feeding on other foods available in the enclosure (e.g., leaves, insects, flowers, seeds, etc.). As it was possible for the monkeys to be hidden from view in the vegetation, records were kept on an interval basis of each monkey's visibility as a part of the behavioral checklist.

Observers were trained in the methods (during adaptation sessions) prior to testing. Reliabilities for behavioral measures (calculated as Σ agree / Σ agree + disagree between two independent observers) were above 90 percent for all behavioral measures: 99 percent for location, an average of 95 percent for all checklist scores, and 92 percent for number and 93 percent for duration of movement bouts per six-minute session. No feeding occurred during event-recorder reliability sessions.

Analysis. Species and sex differences in use of space over ten six-minute time blocks per trial and over trials were assessed using a mixed-model analysis of variance. For this purpose, the enclosure was divided into forty-seven equal-area quadrats, each encompassing twelve locations and totalling 169 m^2. Redundancy in use of space was examined by tabulating the number of different 3.75-m^2 locations entered by each subject over trials and in six-minute time blocks within trials (only the first entry into a location within the time unit of analysis was used for this analysis). Social spacing was assessed using calculations of straight-line interanimal distance for each scoring interval, and by contact and follow scores obtained from checklist data.

Travel patterns were assessed using calculations of distance traveled per fifteen-second interval per subject, and by calculation of straight-line distance traveled in six-minute time blocks (a measure of "effective distance").

Locomotion and feeding data were examined for species, sex, trial, and time-within-trial effects. For this purpose, event-recorder tapes of movement and feeding activity for each subject were scored for frequency and duration of events in ten-minute time blocks (six blocks per trial). Checklist activity (moving, following, feeding, etc.) data were treated in the same manner; summed scores per ten-minute time blocks were used as data for analyses of variance in order to provide a sufficient range of scores (0–40) to approximate a normal distribution.

"Not visible" scores (obtained at the onset of each fifteen-second interval) were less than 8 percent, and analyses of variance indicated no significant effects for this measure for species, sex, trial, or ten-minute block main effects or interactions. Consequently, no "missing data" correction procedures were used in further analyses. A two-tailed α of 0.05 was used throughout for tests of statistical significance.

Results

Use of Space. Both species were highly selective in their use of space. Fig. 7-4 presents schematically the total occupancy scores in each quadrat. A highly clustered distribution is evident for each species. A χ^2 test of pooled scores in the forty-seven quadrats was highly significant for each species ($P < 0.001$, both species) as was the quadrat factor in ANOVA ($F = 5.6; df = 46, 368; P < 0.001$).

Clustering. Although the species were approximately equal in their degree of clustering in this space, it is evident from Fig. 7-4 that they "clustered" in different areas (significant species *x* quadrat interaction, $F = 4.7; df = 46, 368; P < 0.001$). Calculations based on pooled scores by 3.75-m^2 locations indicated a mere 12.7 percent overlap between species in occupancy of all locations. Both species clearly preferred edge and corner locations of whatever surface they were occupying, but *Callicebus* spent most of their time in the aboveground grid system, while *Saimiri* did not. Considerable time was spent at the perimeter of the enclosure by both species, but scores were more widely distributed in *Callicebus* than in *Saimiri*, as indicated by the *Callicebus* scores along the eastern and southern edges of the enclosure and the complete absence of *Saimiri* scores in these areas. *Saimiri* were on the ground more frequently than *Callicebus*; this difference approached significance ($F = 5.12$; $df = 1, 8$; $P < 0.10$). *Saimiri* ground scores would have been much higher had it not been for several subjects' liberal use of the wire (and door hinges in one corner) as travel and resting surfaces.

Travel patterns. Sample hourly *Callicebus* travel patterns are presented in Figs. 7-5, 7-6, and 7-7. Figure 7-5 represents the first trial of male and female for a *Callicebus* pair that occupied an intermediate position for this species in the amount of space covered. The grid area was the clear focus of activity, although on two occasions the female made a foray into the interior, first alone and later accompanied by the male. The female initiated the joint expedition and traveled farther than the male. The fourth trial of this same pair is represented in Fig. 7-6. Again, the grid and feeders are the focus of activity, although several short trips were made to the fence area adjacent to the grid system. As in trial 1, these were all initiated by the

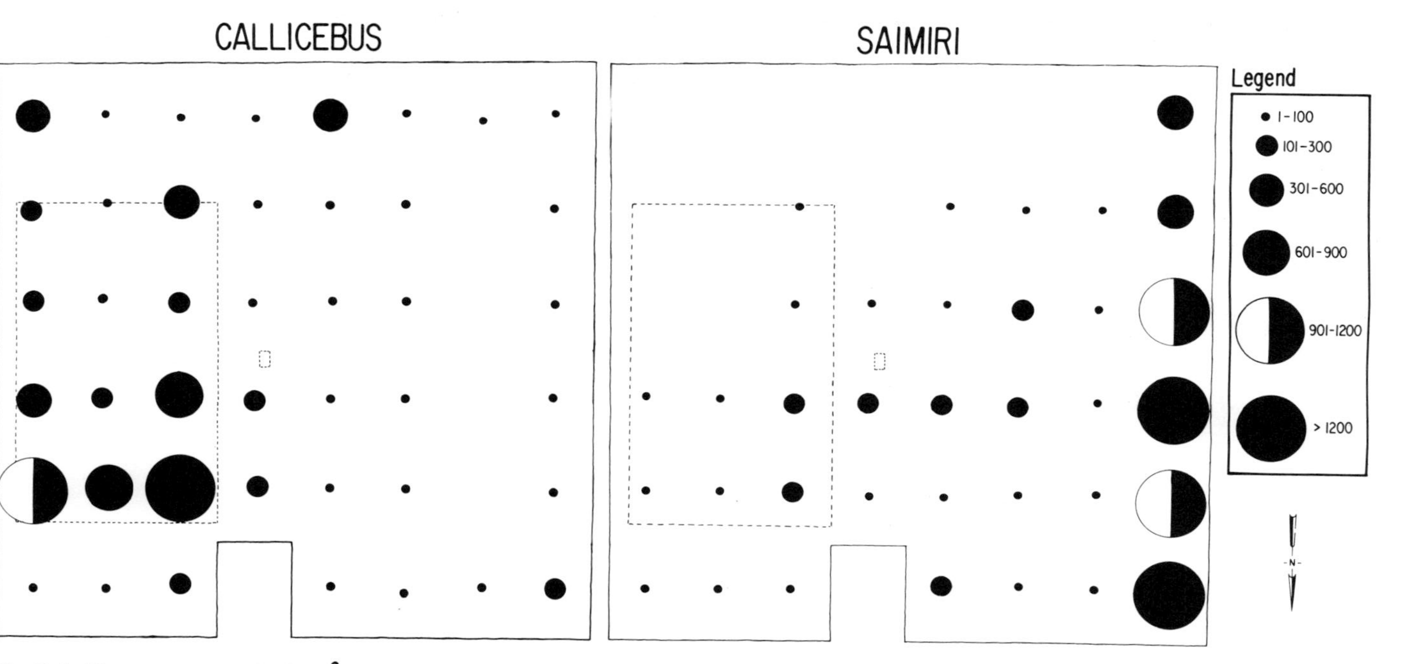

Fig. 7-4. Net occupancy of 169-m^2 quadrats per species over four one-hour trials. Based on point locations of each subject at fifteen-second intervals.

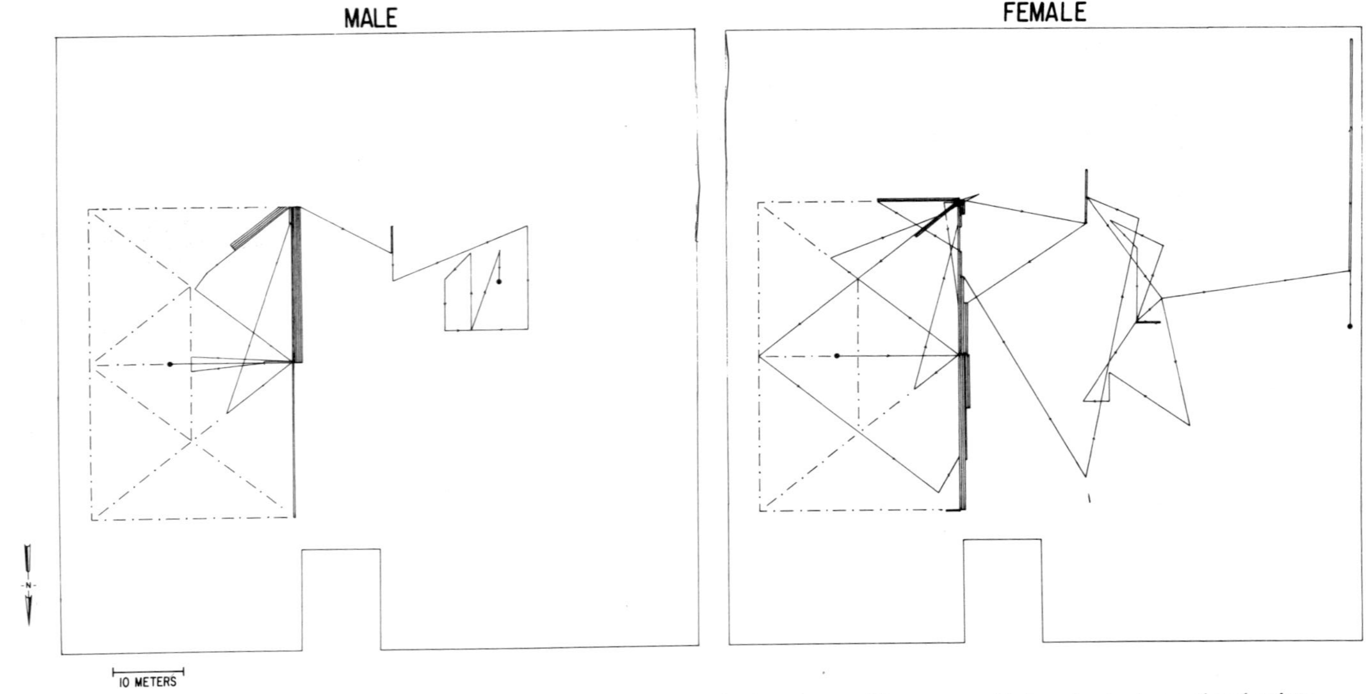

Fig. 7-5. Travel patterns in *Callicebus* pair 1, trial 1 reconstructed from point locations at fifteen-second intervals. ●, start and end points. —, travel sequence. Subjects were released at the center of the above-ground grid system.

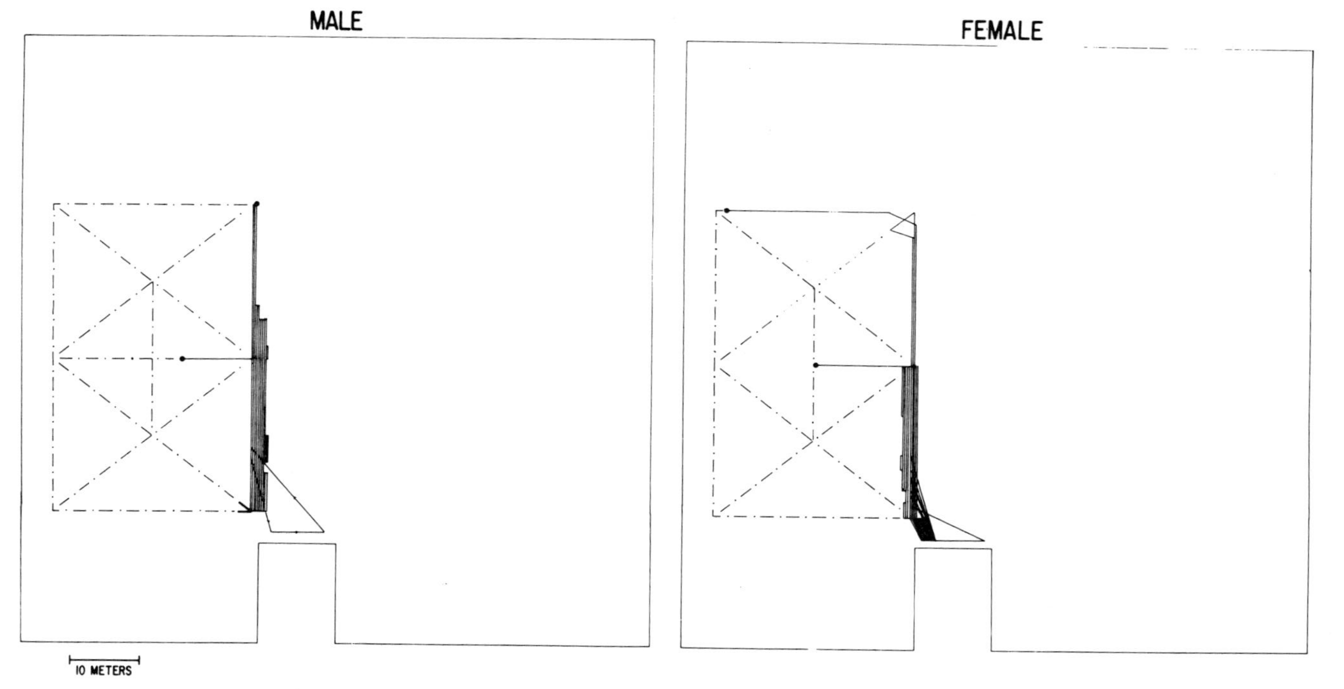

Fig. 7-6. Travel patterns in *Callicebus* pair 1, trial 4 reconstructed from point locations at fifteen-second intervals. ●, start and end points. —, travel sequence. Subjects were released at the center of the above-ground grid system.

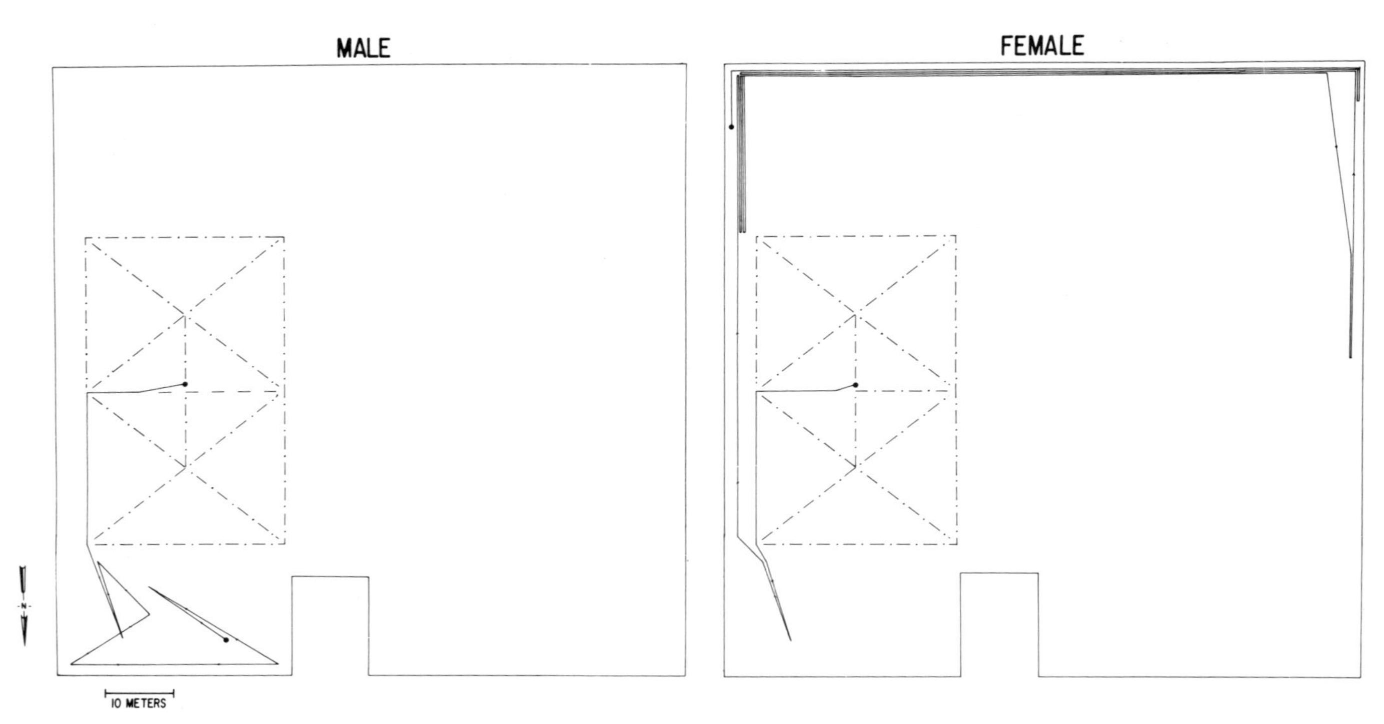

Fig. 7-7. Travel patterns in *Callicebus* pair 5, trial 1 reconstructed from point locations at fifteen-second intervals. •, start and end points. —, travel sequence. Subjects were released at the center of the above-ground grid system.

female. The male followed briefly on one occasion. The female repeatedly returned to the male, which remained at the corner feeder during most of the trial. Lastly, Fig. 7-7 represents travel paths from the first trial of a *Callicebus* pair that was atypical in that it spent most of each trial away from the grid system. As in the previous pair, the female is the more active animal, ranging far along two sides of the enclosure. A second atypical feature of this pair's travel pattern is the low degree of coordination between the pairmates (contrast Fig. 7-7 with Figs. 7-5 and 7-6). Although the pairmates left the grid area together, the female resumed activity within a few minutes, while the male did not.

Turning now to *Saimiri*, Figs. 7-8 and 7-9 present sample travel patterns of a typical pair. Figure 7-8 is first-trial data for male and female; Fig. 7-9 is fourth-trial data. Activity for both animals was entirely restricted to one edge of the enclosure during the first trial. During the fourth trial, the male returned in mid-trial and end-trial to the observer platform and grid area, both times coming through the cage interior rather than along the fence edge. The female's pattern is relatively unchanged; there were no trips into the interior except to get to the western fence at the start of the trial.

Temporal patterns. Clear species contrasts in use of space over time emerge when changes over trials in the two species are compared (see Fig. 7-10). In trial 1, *Callicebus* entered slightly more locations than *Saimiri* ($\bar{X} = 33.2$, *Callicebus* versus 25.9, *Saimiri*, n.s., Mann-Whitney *U*). Trial 2 means were quite close (25.7, *Callicebus*, versus 28.6, *Saimiri*). However, in trials 3 and 4, *Saimiri* means were clearly higher than those of *Callicebus* (e.g., 15.4, *Callicebus* versus 37.3, *Saimiri*, trial 4; $P < 0.02$, Mann-Whitney *U*). All ten *Callicebus* subjects entered fewer locations on trial 4 than trial 1 (mean individual decline = 54 percent; $P < 0.01$, Wilcoxon matched-pairs signed-ranks). Five of ten *Saimiri* subjects entered more locations, and the mean individual increase was 44 percent (n.s., Wilcoxon test).

Sex differences. Within species, the sexes varied in opposite directions in their use of space: females' scores decreased more than males' in four of five *Callicebus* pairs, but increased more or decreased less than males' in four of five *Saimiri* pairs (see Fig. 7-10).

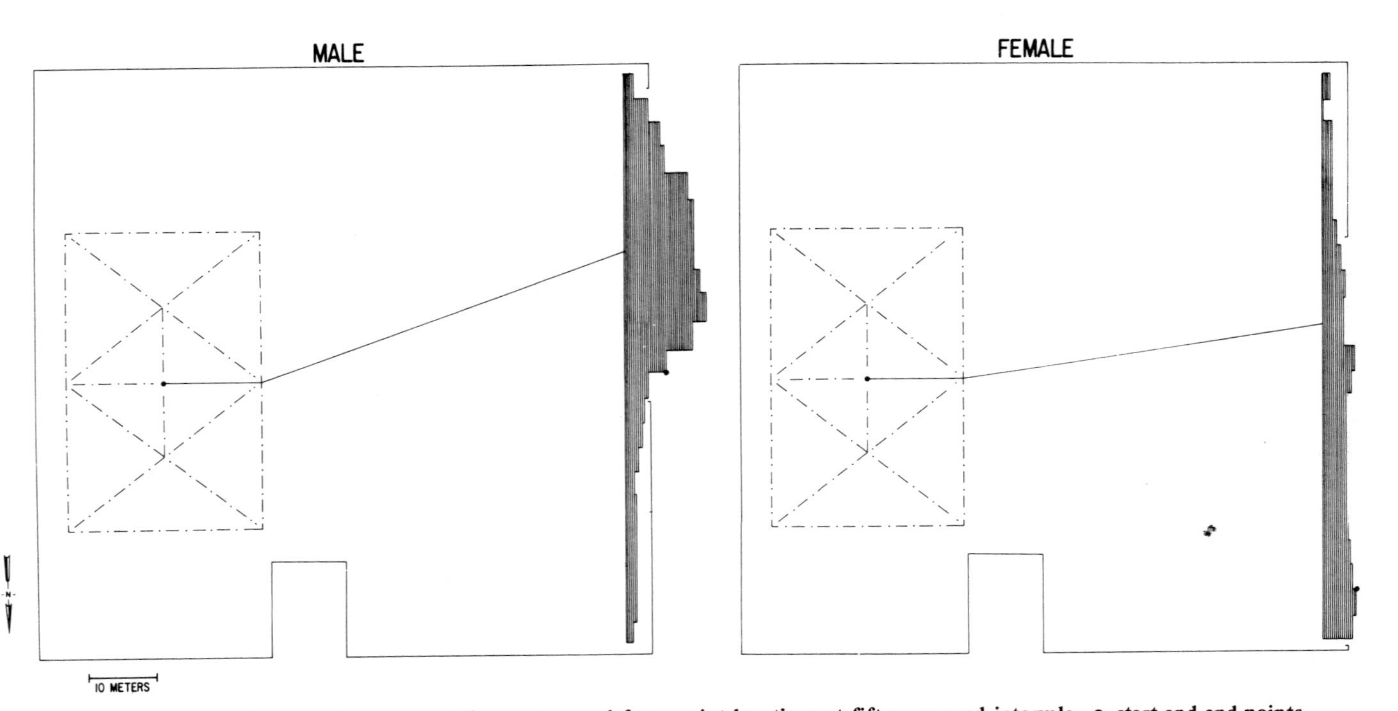

Fig. 7-8. Travel patterns in *Saimiri* pair 5, trial 2 reconstructed from point locations at fifteen-second intervals. ●, start and end points. —, travel sequence. Subjects were released at the center of the above-ground grid system.

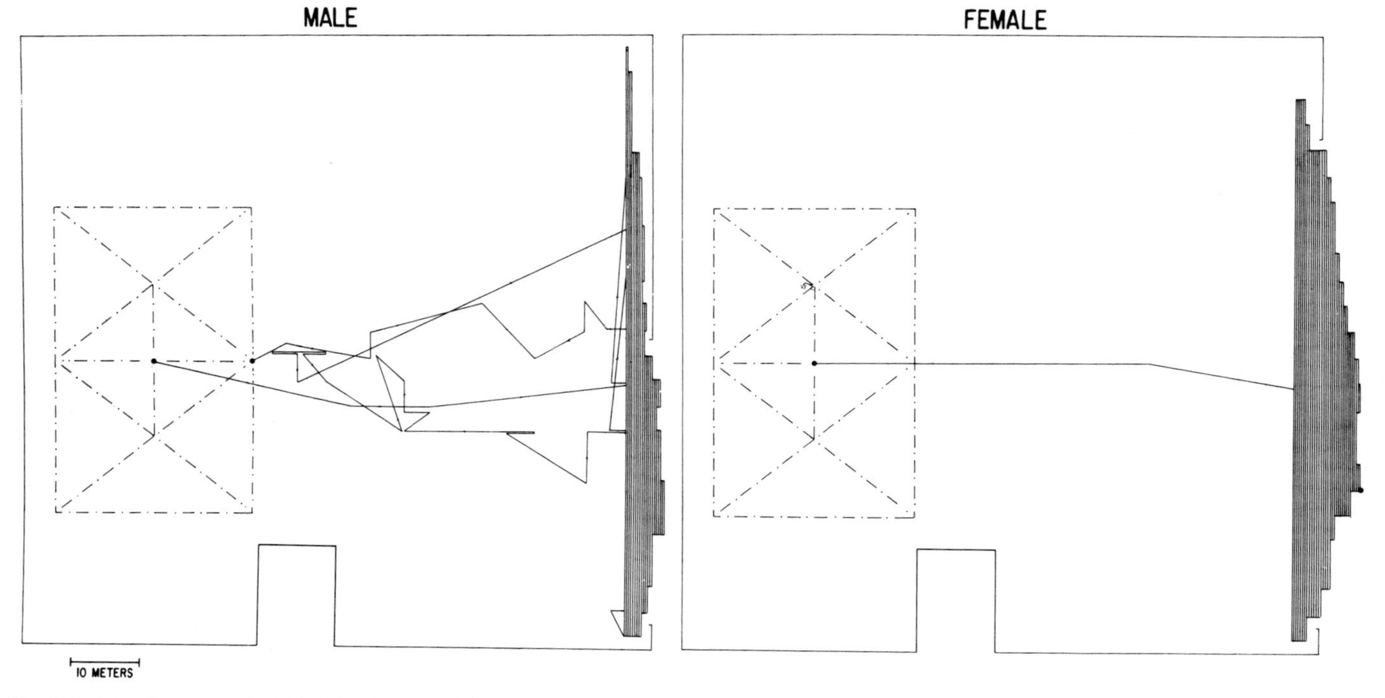

Fig. 7-9. Travel patterns in *Saimiri* pair 5, trial 4 reconstructed from point locations at fifteen-second intervals. •, start and end points. —, travel sequence. Subjects were released at the center of the above-ground grid system.

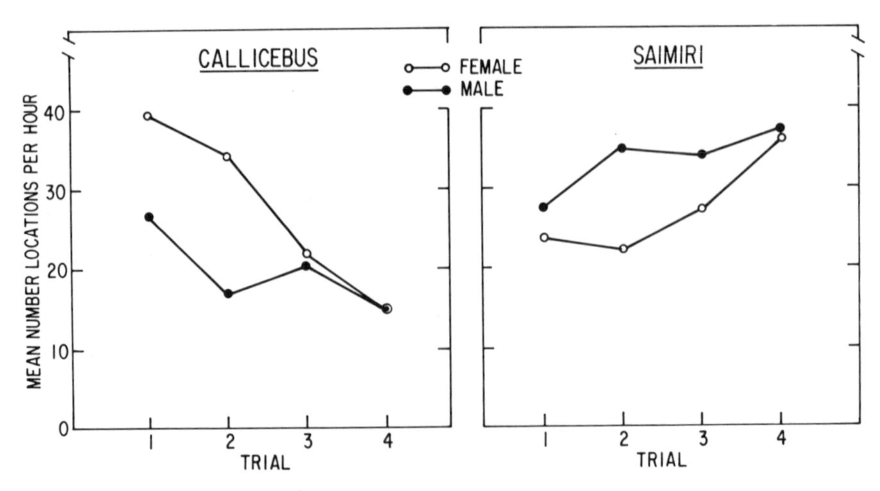

Fig. 7-10. Number of 3.75-m^2 locations entered per trial calculated from fifteen-second interval data. Maximum of one score per location per trial.

Callicebus females entered more locations than males in three of five pairs in trial 1 (93 percent more, average), and the overall female mean was substantially higher (39.6, female versus 26.8, male; see Fig. 7-10). Examination of male and female scores in the ten least used quadrats in which occupancy occurred reveals that females used seven of them more than males on the first trial. However, female location entries declined steeply over trials; by trial 4, male and female means were identical. Similarly, female use of all of the ten least used quadrats was greater than male use in trial 1, but was no different on trial 4. The pattern was reversed in *Saimiri*, where the mean number of locations entered per female increased by 55 percent over trials 1 to 4, a larger increase than in males (35 percent), and the magnitude of male-female differences was inconsistent. However, females expanded their travel activities into areas which were already being used to a moderate degree. This is indicated by the data on the use of the ten least frequently occupied quadrats: males were the only users of six of these ten quadrats, and the more frequent users of seven of the ten. Furthermore, male use of these quadrats was spread out over the hour, 40 percent of the use occurring after the first six minutes, whereas female use occurred exclusively in the first six minutes. In most cases female use could be attributed to the run from the release point to the western fence.

Changes in preferred locations. The species also differed in the way their use of preferred locations changed over time (see Fig. 7-11). Titis' use of their ten most preferred quadrats increased steadily over trials 1 to 4 (from 49 percent to 89 percent of all intervals) whereas squirrel monkeys' use of their ten most preferred quadrats was consistently high over all trials, averaging 90 percent of all intervals (n.s., χ^2). The trends in use of least preferred quadrats were in the opposite direction: titis' use of least preferred quadrats declined steadily over trials, while squirrel monkeys' usage was more consistent ($P <$ 0.001, χ^2). For example, only 15 percent of titis' use of these quadrats occurred in trials 3 and 4; the comparable score for squirrel monkeys is 29 percent.

In spite of the fact that squirrel monkeys entered more locations, the species did not differ in the tendency to visit the same location on successive trials. Repeated visits or redundancy in individual use of space across trials was measured by the mean proportion of total locations entered over four trials which were entered on each trial. That the species were nearly equivalent on this measure of space use is indicated in Fig. 7-12 (*Callicebus* $\bar{X} = 0.49$; *Saimiri* $\bar{X} = 0.47$).

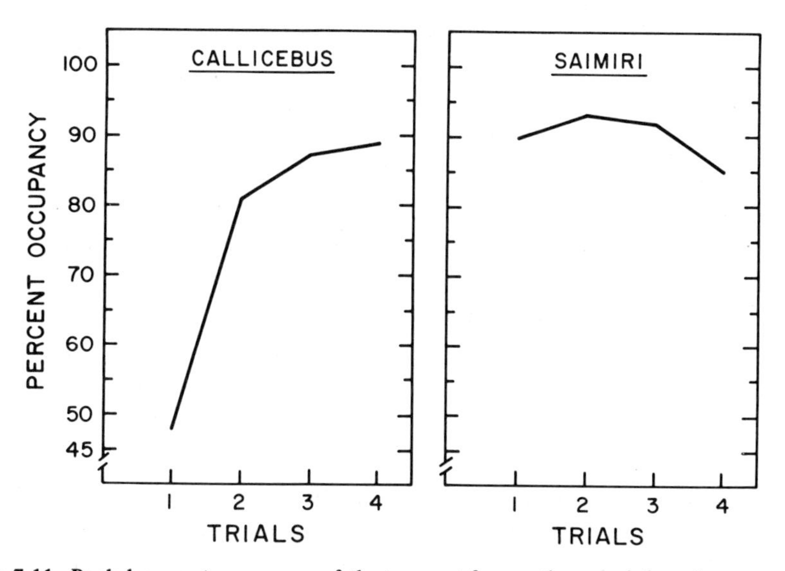

Fig. 7-11. Pooled percent occupancy of the ten most frequently occupied quadrats per species over four one-hour trials. Based on point locations of each subject at fifteen-second intervals.

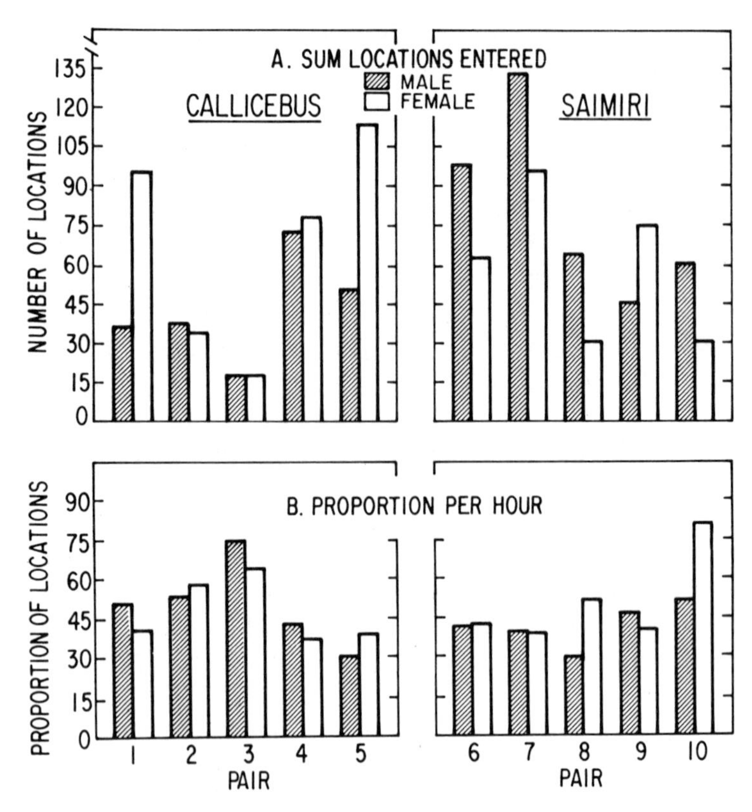

Fig. 7-12. Mean number and mean proportion of all locations entered which were occupied on each trial. Maximum of one score per location per trial per subject.

Temporal changes in location entry within trials indicate a significant decline in exploratory travel within the hour in both sexes of both species (runs test on ranks, $P < 0.05$, all sexes and species, see Fig. 7-13). The decline in number of locations entered is proportionally greater in *Callicebus* (species $\bar{X}$ of 7.4, block 1, to 3.8, block 10, or 49 percent decline; versus 10.6 to 6.2, or 41 percent decline in *Saimiri*), and is greater in males than females of both species.

Patterns of exploration and locomotion. Qualitative species differences in patterns of locomotion and exploration are of interest. Titis traveling across the ground—especially open areas—sometimes used a bounding gait which propelled them forward and upward at relatively high velocity. The apex of the leap often occurred nearly a

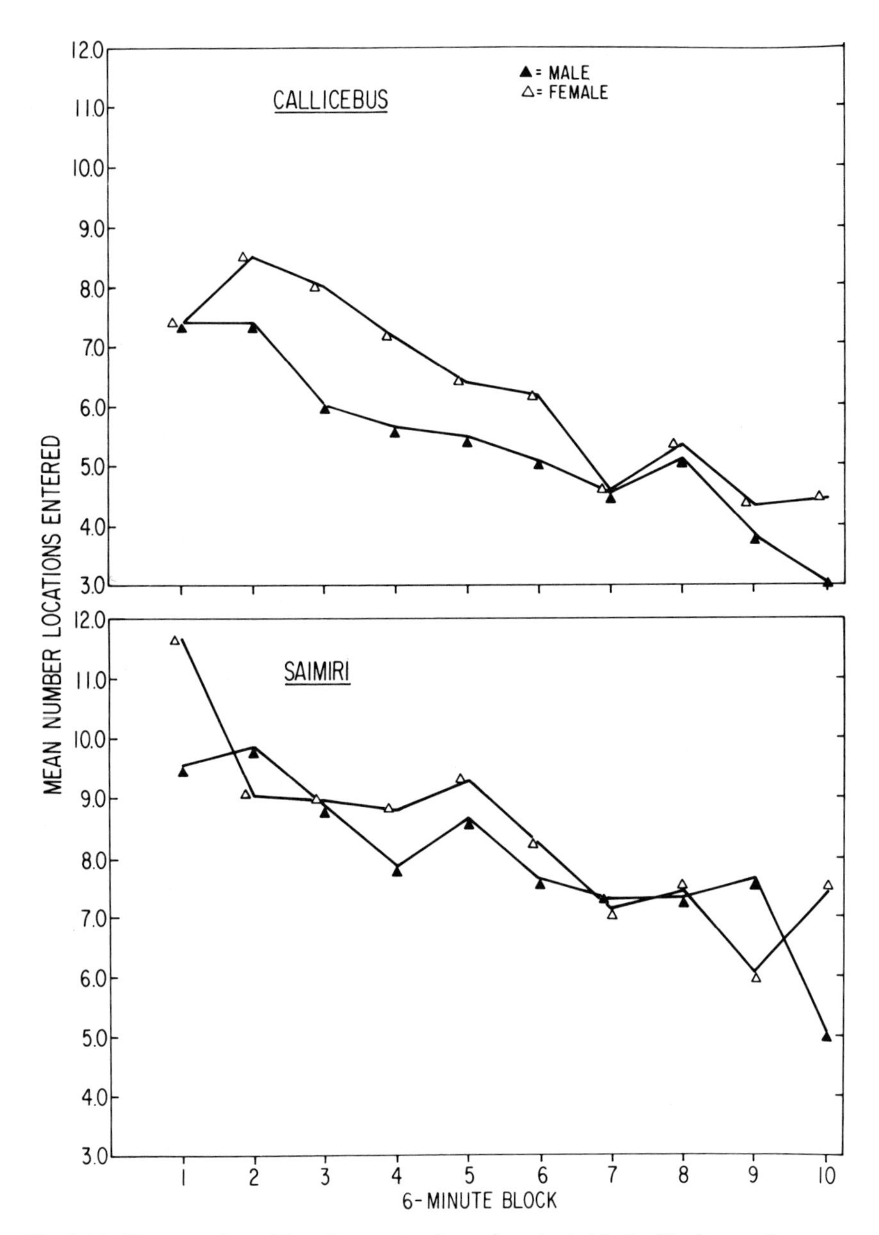

Fig. 7-13. Mean number of locations entered per six-minute block. Maximum of one score per location per six-minute blocks.

meter above the ground, and forward distance was often 1.5 m or more. This seemed to be entirely produced by leg thrusting. Squirrel monkeys never "bounded" as high as did the titis, although they did bound in high grass. They usually ran rapidly using a two-beat flat-footed diagonal gait or a more graceful two-beat bounding gait. Both species traveled equally rapidly at top observed speed (about as fast as a normally athletic observer could run in short bursts through high grass).

Squirrel monkeys spent much time climbing high on the wire, and moved on it rapidly and with agility in both horizontal and vertical planes (head up or down). One squirrel monkey male climbed up a structural pole in midfield and spent sixteen minutes upside down on the roof wire. He descended by voluntarily dropping the 10 m from roof to ground. Other monkeys also made use of these poles alongside the fence to climb up towards the roof. One female leapt back and forth from pole to fence several times within a few seconds, apparently chasing something (unseen) of great interest. Small trees were also climbed, as were small poles. Variability in locomotor patterns as well as use of a variety of travel surfaces was evident within as well as among individuals.

Titis were much more conservative in their locomotor patterns. Only one pair showed much inclination to climb the wire (the pair whose travel patterns are sketched in Figs. 7-5 and 7-6), and they were much slower doing so than *Saimiri*. *Callicebus* did not climb (or even attempt to climb) a pole. A female once leapt into a small tree and then immediately out again; this constituted the only use of a surface other than the lumber grid, ground, or wire. Overall, these differences are in line with Erickson's (1963) categorizations of *Callicebus* as "springers" and *Saimiri* as "climbers."

Postural patterns. A final qualitative difference in locomotor style concerns postural patterns. *Saimiri* (particularly males) occasionally stood bipedally and visually scanned the surroundings. This occurred most frequently when on the ground in the cage interior. *Callicebus* never did this; in fact, they seemed singularly inattentive (visually, at least) to distant surroundings. For example, a passing cyclist was the object of prolonged visual scrutiny by a *Saimiri* pair. *Callicebus* did not respond with overt head orientation under similar circumstances (passing automobiles or cyclists).

Locomotor Activity. Patterns of locomotor activity differed substantially between the two species. Squirrel monkeys were more active, as indicated in several related measures (see Fig. 7-14). Species differences were significant for the number of intervals in which movement was scored [individual $\bar{X}$ score per trial was 102.4 (43 percent of all intervals) *Saimiri*; 53.4 (22 percent of all intervals) *Callicebus*; $F = 9.8$; $df = 1, 8$; $P < 0.025$], and in proportion of time spent moving (individual $\bar{X}$ proportion per trial was 0.21, for *Callicebus*; 0.39 for *Saimiri*; $F = 8.5$; $df = 1, 8$; $P < 0.025$). Individual variability within species overshadowed significant species effects in absolute frequency and duration of movement, although *Saimiri* individual unweighted trial means were substantially higher in both cases ($\bar{X}$ absolute frequency was 106.9, *Saimiri* versus 87.1, *Callicebus*; $\bar{X}$ bout duration was 16.5 seconds, *Saimiri* versus 9.8, *Callicebus*, see Fig. 7-14). In addition to traveling more frequently and for longer durations per bout, *Saimiri* traveled 88 percent farther per hour than did *Callicebus* (*Saimiri* $\bar{X} = 1{,}049$ m/hour; *Callicebus* $\bar{X} = 559$; Mann-Whitney U, $P < 0.02$; see Figs. 7-15 and 7-16).

Sex differences. As in location entry, the sexes differed in each species on all four measures of locomotor activity. The patterns were not the same in the two species, however. Although titi females generally entered more locations than male pairmates (see Fig. 7-8), they moved less often and for shorter durations ($\bar{X}$ absolute frequency per trial was 17.6 versus 21.4; $\bar{X}$ intervals per trial were 46.5 versus 60.4; $\bar{X}$ proportion time spent moving was 0.18 versus 0.25; $\bar{X}$ bout duration was 10.2 seconds versus 12.5 seconds, all female to male, respectively). The direction of sex differences was reversed in squirrel monkeys: individual female trial means were higher on three measures (113.2 versus 91.7 $\bar{X}$ intervals per trial; 0.41 versus 0.36 $\bar{X}$ proportion time spent moving; 15.5 seconds versus 13.5 seconds, $\bar{X}$ bout duration, all female to male, respectively) and nearly equivalent in the last measure, absolute frequency (17.6 female, versus 18.0, male).

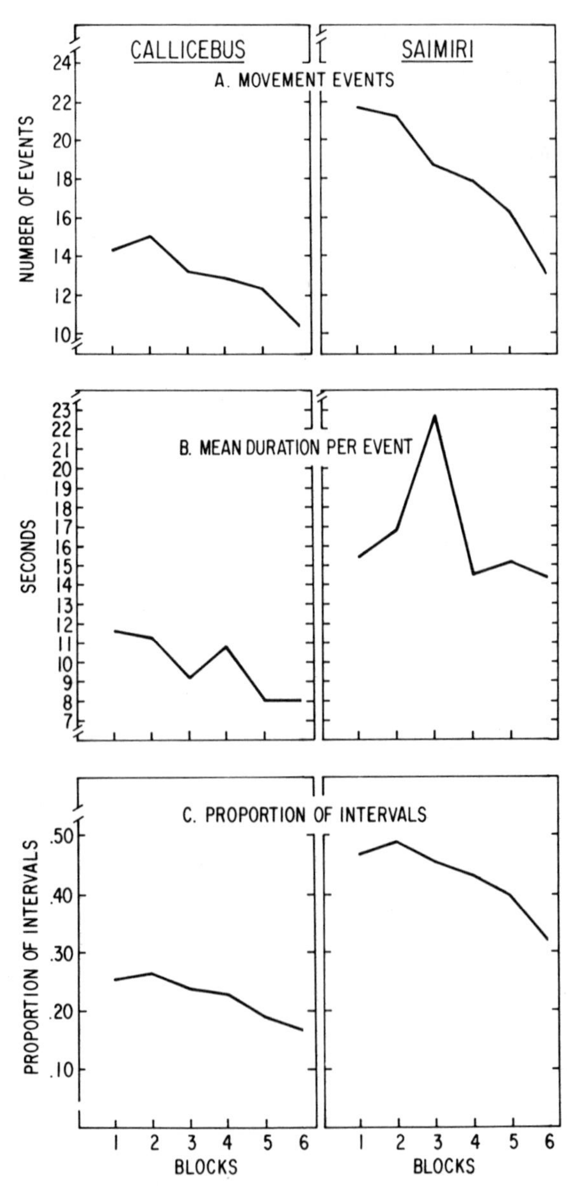

Fig. 7-14. Locomotor activity per ten-minute block. (a) Locomotion events taken from event-recorder data. Each event was ⩾ one second and occurred after ⩾ three seconds of locomotor inactivity. (b) Mean duration per bout calculated from event-recorder data. (c) Proportion of fifteen-second intervals on which locomotion was scored.

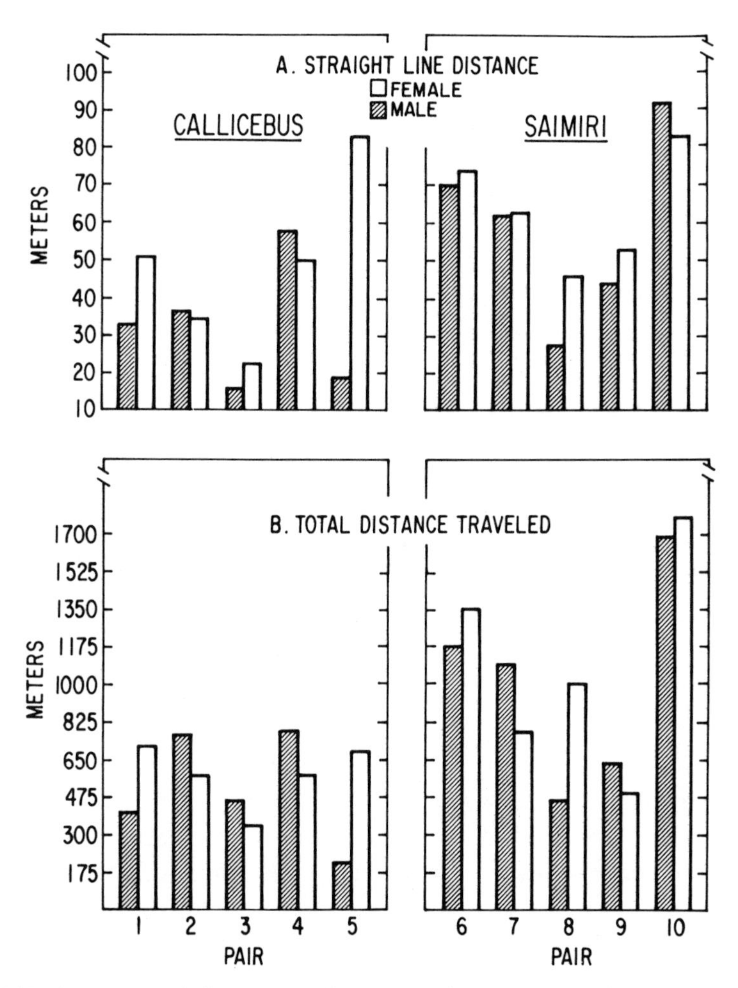

Fig. 7-15. Average travel distances per hour per subject. (a) Sum of distances between point locations at onset and termination of six-minute periods. (b) Sum of distances between point locations at onset and termination of fifteen-second intervals.

Directness of travel. A rather different index of locomotor activity is obtained by comparing directness of travel. This was obtained by calculating correlation coefficients (Pearson product moment) between effective (straight-line) distance and pathlength. Straight-line distance scores for six-minute blocks were matched with scores for total distance moved over the same six-minute period, providing a set of forty paired scores for each subject (ten six-minute blocks per

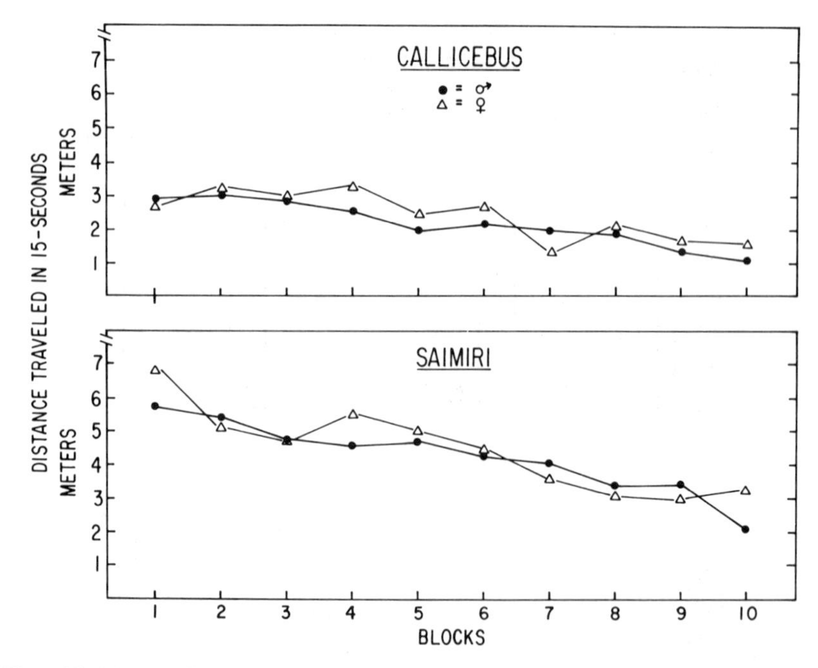

Fig. 7-16. Average distance moved in fifteen seconds per six-minute blocks based on point location data.

trial X four trials). The species did not differ significantly in directness of travel according to this measure (*Callicebus* $\overline{X}$ correlation was +0.51, range was +0.25 to +0.80; *Saimiri* $\overline{X}$ correlation was +0.42, range was +0.09 to +0.62). Spearman rank-order correlations by individuals on numbers of locations entered and straight-line distance traveled per hour were calculated to assess the possibility that the amount of space used was correlated with directness of travel. Correlations were substantial in both species: +0.77, *Callicebus*, and +0.46, *Saimiri.* In general, then, more direct travel was associated with a more diverse use of space within individuals in each species. However, across species, even though more locations were entered by squirrel monkeys than by titis, average individual "directness" was somewhat lower for squirrel monkeys.

Distance traveled. Distance traveled per hour declined over trials in both species, although by a larger percentage in *Callicebus* than in *Saimiri* (32 percent versus 11 percent). Mean scores for trials 1 and 4 were 632 m/hour and 431 m/hour, *Callicebus*; 1,076 m/hour and 957 m/hour, *Saimiri*. Comparisons of trials 1 and 4 indicated a significant decline in *Callicebus* ($P < 0.02$), but not in *Saimiri* (Wilcoxon matched-pair signed-ranks). Species differences in distance traveled were not significant in trials 1 and 2, but were significant in trials 3 and 4 ($P < 0.02$, both cases, Mann-Whitney U).

Effective distance of travel. Changes in effective distance over trials were clear in both species; however, the direction of change was different. Straight-line distance declined significantly over trials 1 to 4 in titis (57 percent decline; $\bar{X} = 54.7$ m/hour versus 23.6 m/hour, respectively, $P < 0.01$, Wilcoxon matched-pairs signed-ranks). This is a proportionally larger decline than was evidenced in total distance traveled, indicating that travel paths became more redundant as movement decreased. In squirrel monkeys, by contrast, mean straight-line distance per hour increased from trials 1 to 4 ($\bar{X} = 51.1$ m/hour to 62.8 m/hour, n.s.), even though total distance traveled declined slightly. These data suggest less redundancy in travel on the fourth trial in squirrel monkeys.

With respect to changes over trials in total travel distance, sex differences were greater in titis than in squirrel monkeys. In titis, changes were greater in females than males. The average female score declined from 672 m/hour in trial 1 to 381 m/hour in trial 4, a 43 percent decrease ($P < 0.02$, dependent t); comparable male scores were 558 m/hour and 458 m/hour, a 22 percent decrease (n.s.). In squirrel monkeys, changes over trials 1 and 4 were slight and nonsignificant in both sexes (17 percent decline in males, 1062 m/hour to 882 m/hour; 5 percent decline in females, 1033 m/hour to 979 m/hour). Sex differences in straight-line distance changes over trials were similar to those in total distance traveled: titi females showed greater declines than males (68 percent; $P < 0.10$, versus 37 percent, n.s., dependent t), but squirrel monkeys of both sexes showed equivalent increases (26 percent, males and 20 percent, females, n.s.).

Temporal aspects of activity. Within trials, the temporal dimension was a potent factor in locomotor activity, just as it was for location preference and number of locations entered. This is clearly indicated in Fig. 7-14, which shows that individual mean proportions of intervals in which movement occurred in ten-minute blocks declined by nearly one-third in each species (from 0.47 to 0.32 for *Saimiri*, and 0.25 to 0.17, *Callicebus*). The pattern was nearly identical for the two species: each showed a slight increase in block 2, and steady decreases in subsequent blocks. Similar decreases over time were apparent in other measures of locomotor activity, and block effects were significant for three of these: distance moved per interval (six-minute blocks, $F = 7.4$; $df = 9, 72$; $P < 0.001$; see Fig. 7-16); proportion of time spent moving (ten-minute blocks, $F = 5.6$, $df = 5, 40$; $P < 0.001$); and absolute frequency of movement (ten-minute blocks, $F = 8.7$; $df = 5, 40$; $P < 0.001$). The mean duration of movement bouts declined also, although this effect was not significant for either species.

Summary. In summary, temporal changes in travel activities varied in the species, and varied in the sexes in different ways in each species. Declining travel was accompanied by a more restricted use of space in titis, but not in squirrel monkeys. Over trials, titis' movements became less frequent and more restrictive (fewer locations entered, straight-line and total distance declined). Titi females altered their activities over time more than males, both within and across trials. Squirrel monkeys' travel activities, although declining within trials, became less restricted over trials (more locations entered, straight-line distance increased, distance traveled about the same). Squirrel monkey females showed changes over time within and across trials equivalent to those in males. In contrast to titis, where females moved less than males, squirrel monkey females moved as frequently, in longer bouts, and over greater distances than did males.

Social Aspects of Use of Space. As expected, social aspects of use of space differed considerably in the two species.

Interanimal distance. Because of an extreme score on one trial for *Callicebus*, species differences were not significant in parametric analyses of interanimal distance. Nevertheless, *Saimiri* had much higher average interanimal distance scores (*Callicebus* $\overline{X} = 10.2$ m; *Saimiri* $\overline{X} = 18.1$ m; see Fig. 7-17). Fisher's exact probabilities for distribu-

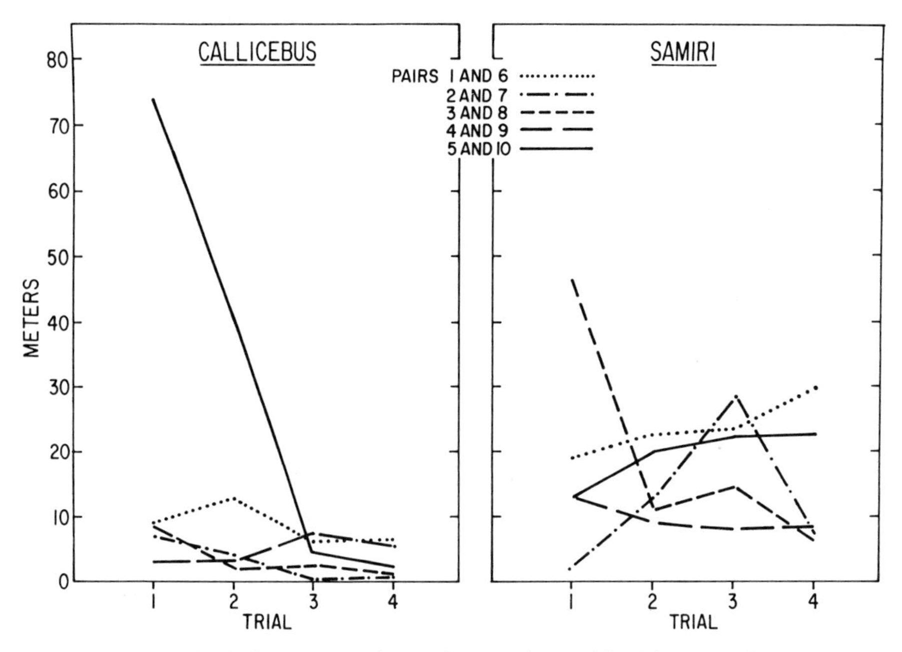

Fig. 7-17. Interanimal distance per pair over four one-hour trials. Distances calculated from simultaneous location sampling at fifteen-second intervals.

tion of scores around the median indicated that the average distance between *Saimiri* pairmates was greater than that for *Callicebus* pairmates overall ($P = 0.10$, by pair); and approached or reached significance in three out of four trials ($P = 0.004$, trials 3 and 4; and $P = 0.10$, trial 1). Looking at the data in a different way, although there was considerable overlap between species in the average *maximum* interanimal distance per six-minute block per pair, the average *minimum* distance score per six-minute block was less than 1 m for four out of five titi pairs, but at least 2.7 m for any squirrel monkey pair. Titi pairs were in the same 3.75-m^2 location at least once in 85.5 percent of all six-minute blocks; the comparable figure for *Saimiri* is 75.0 percent. In other words, titi pairmates nearly always approached each other quite closely at least once every six minutes, while squirrel monkey pairmates did so less often, even though use of space overall was extremely similar for pairmates of both species ($\overline{X}$ Kendall's tau on quadrat ranks in pairmates was +0.79, *Callicebus*; +0.68, *Saimiri*; $P < 0.01$ for all pairs, both species).

Following responses. Data on following reflect a more active individual role in the maintenance of social proximity in titis (see Fig. 7-18). Titis followed the pairmate on significantly more of all moves scored than did squirrel monkeys (mean percent moves which were also follows was 18.5, *Callicebus*, versus 6.6, *Saimiri*; Mann-Whitney $U = 23.5$; $P \leqslant 0.05$). Most of the following in titi pairs was done by females (female $\bar{X} = 30.1$ percent, male $\bar{X} = 7.0$ percent; female > male, all five pairs). The reverse was true in squirrel monkeys: females followed less often than males in four out of five pairs (female $\bar{X} = 4.3$ percent; male $\bar{X} = 8.8$ percent). Thus, in both species, the pairmate of the more frequently active sex (male in *Callicebus*, female in *Saimiri*) was followed more often than it followed. Most squirrel monkey follows occurred along the western fence, which, of course, was the locus of most squirrel monkey activity, while most titi follows occurred along the perimeter of the grid (see Fig. 7-19). Observers felt that follows by titis were nearly always intentional;

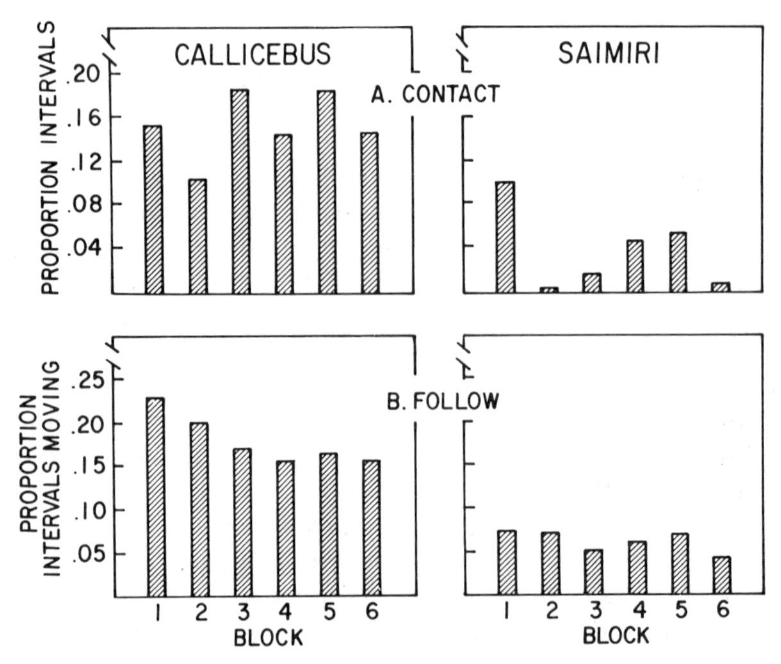

Fig. 7-18. Average social orientation measures per ten-minute block. (a) Mean proportion per pair of fifteen-second intervals on which social contact was scored. (b) Mean proportion of intervals per subject on which locomotion was scored for which following was also scored.

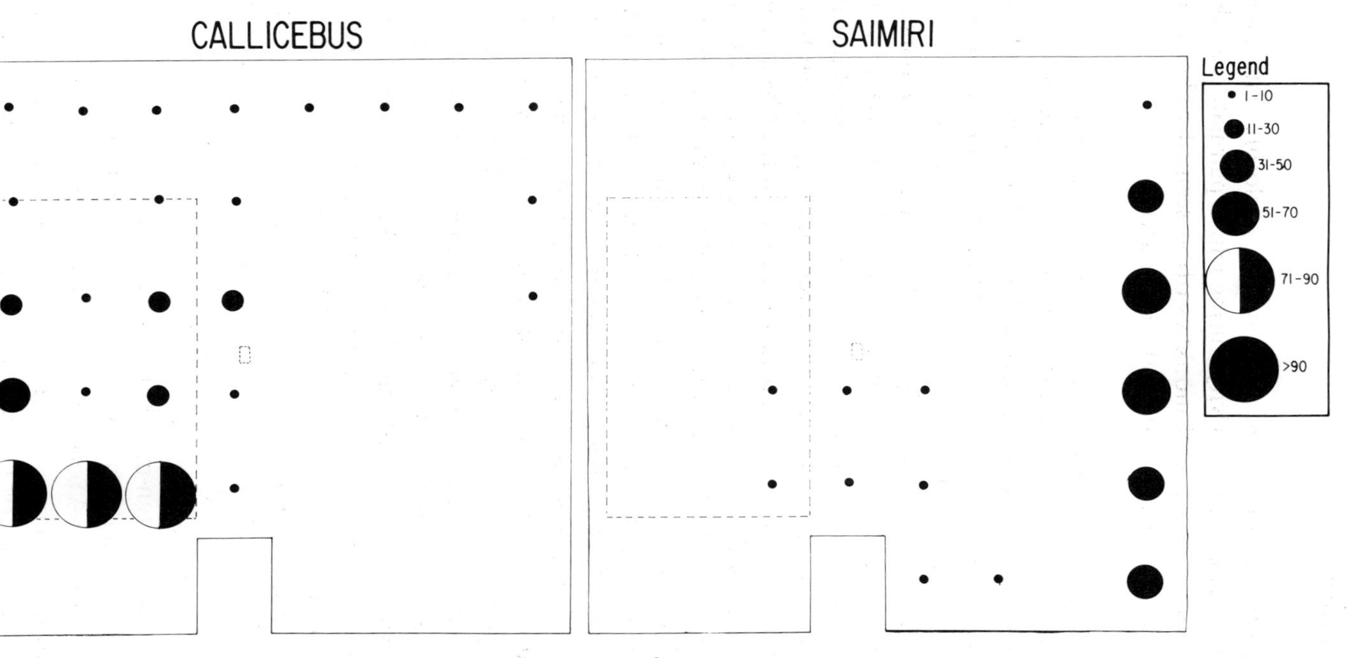

Fig. 7-19. Net frequency of following per 169-m^2 quadrat based on fifteen-second interval data.

some occurred in little-used areas of the enclosure, and many occurred after periods of inactivity during which pairmates sat close together. Follows by squirrel monkeys often seemed to be coincidental results of frequent activity in the same area.

Social contact. Contact data provide additional information on species differences in social proximity. Titis scored many more contacts than did squirrel monkeys (individual $\bar{X}$ of 15 percent of all possible intervals for *Callicebus* versus 2 percent for *Saimiri*; see Fig. 7-18). Because of an extremely high score by one *Saimiri* pair on one trial, species differences were not significant by analysis of variance. With the exception of this pair, all titi pairs scored more contacts than did squirrel monkey pairs. One squirrel monkey pair scored no contacts at all over four trials; two others scored contacts on one trial only. No contacts were scored on a total of thirteen out of twenty squirrel monkey trials, while three or more contacts were scored for every titi trial. *Saimiri* contacts occurred principally in small areas of high-intensity use (see Fig. 7-20). Excluding the extreme trial data on the single *Saimiri* pair (seventy-three contacts in one trial), there was a total of thirty-four contacts. Of this number, 47 percent occurred near one feeder station and was attributable to a single pair; 38 percent (thirteen of thirty-four) occurred in the northwest corner, and only 15 percent (five of thirty-four) occurred along the western fence. A large percentage of squirrel monkey contacts (32 percent) occurred as the animals climbed on the fence wire. A total of 714 contacts were scored for titi monkeys. Like following, contacts occurred in a variety of locations: 199 occurred along the fence, 2 occurred in an interior location, and 513 occurred around the grid area (see Fig. 7-20). Of these contacts 28 percent (202) occurred on the ground (99 percent, or 199 of these were scored by one pair); the rest were on the grid. Frequency of contacts remained relatively constant in titis, ranging between 10.6 percent and 17.6 percent of all intervals in each ten-minute block. In squirrel monkeys, there was a clear peak in the first ten minutes (10.0 percent in block 1; no other block above 5.4 percent). This is in contrast to the time course of proportions of moves which were follows, which declined steadily over the hour for titis but remained stable in squirrel monkeys (see Fig. 7-18).

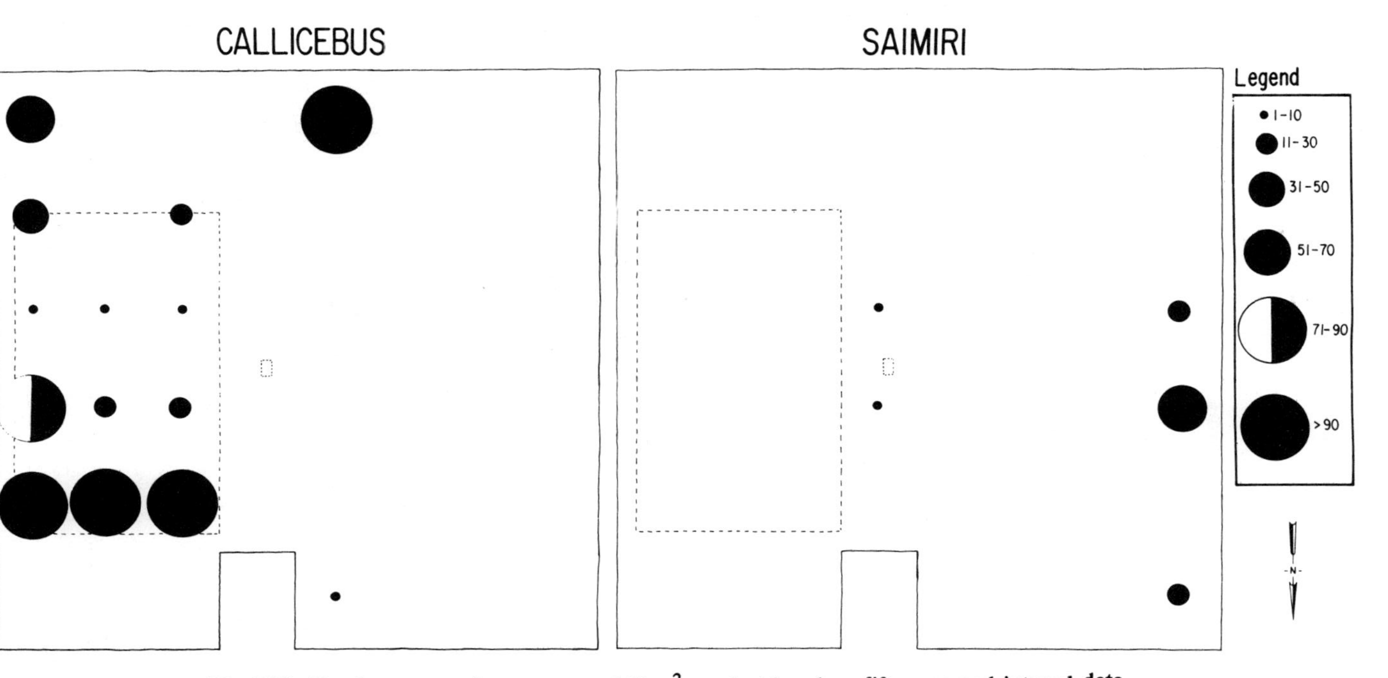

Fig. 7-20. Net frequency of contacts per 169-m^2 quadrat based on fifteen-second interval data.

Spatial coordination. There are several sources of evidence that females were primarily responsible for the close spatial coordination observed in titi pairs. Analyses of relationships between contact and following scores indicated a moderate positive correlation between female following and contact scores over five pairs ($r_s = +0.30$). The male correlation, in contrast, was -0.50. Similarly, the correlation between proportions of male and female moves that were follows was moderately negative ($r_s = -0.20$), suggesting that actively following females were not followed to the same degree by their pairmates. Comparable analyses of squirrel monkey pairs indicated no clear pattern of male and female differences. For example, frequency of contacts and move/follow proportions were essentially uncorrelated in both sexes ($r_s = -0.10$ for both).

Activity versus contact. A second correlational question concerning social orientation is whether contact is reduced when either or both pairmates are relatively active travelers or feeders. Again, the species differ quite clearly: titis show low positive correlations between intervals moving and in contact ($r_x = +0.40$, females and $+0.10$, males); squirrel monkeys of both sexes show substantial negative correlations for the same variables (-0.90, females and -0.60, males). Thus, generally active titis maintained moderately high levels of social contacts, whereas the reverse is true for squirrel monkeys.

Feeding. Feeding activities were organized quite differently in the two species on every dimension considered: location, utilized resources, temporal pattern, and social orientation. We begin with an examination of where feeding took place: *Callicebus* fed almost exclusively on the grid system. Less than 1 percent of intervals feeding were scored off the grid in this species. Although *Saimiri* also fed quite often in feeders, a substantial proportion of feeding took place off the grid in this species (42 percent of all intervals feeding) (see Fig. 7-21).

Location of feeding. As might be expected from the location of feeding activities, titi monkeys fed principally on foods obtained from the feeder boxes (97 percent of all feeding intervals), while squirrel monkeys fed on nonprovisioned foods on a substantial pro-

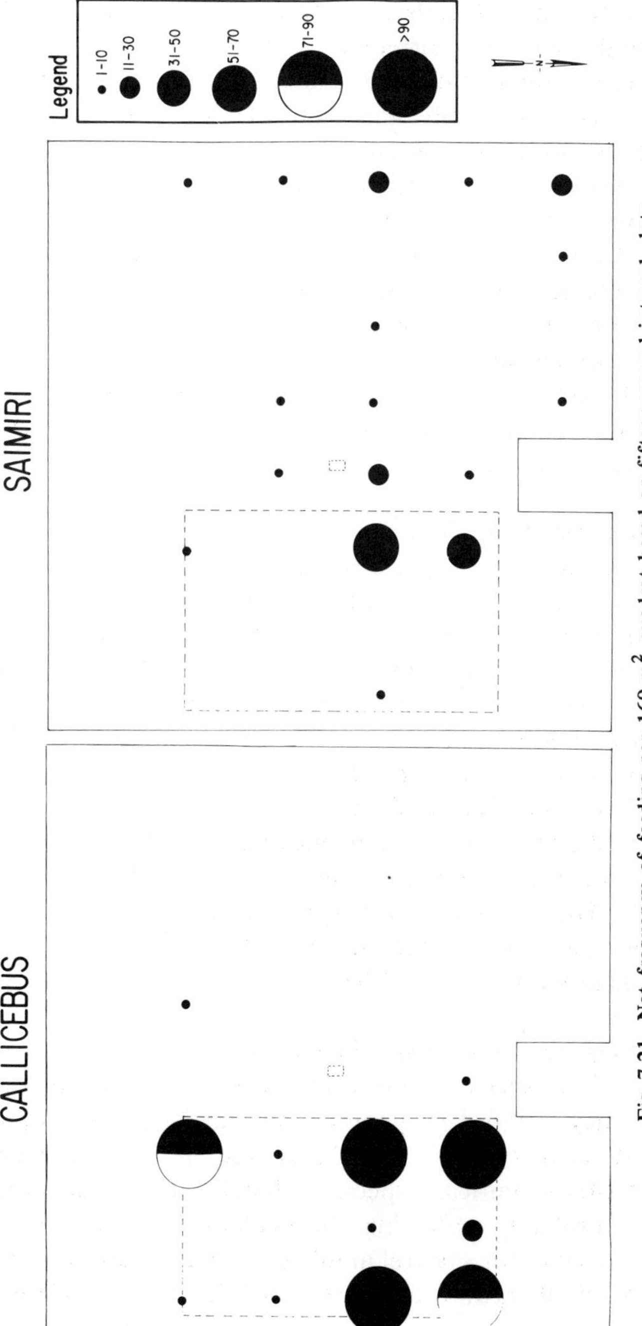

Fig. 7-21. Net frequency of feeding per 169-m² quadrat based on fifteen-second interval data.

portion of feeding intervals (36 percent). These data underscore a fundamental qualitative difference in feeding style. Squirrel monkeys are active, opportunistic foragers. In the one-hectare enclosure, squirrel monkeys carefully inspected and found food items in sunflower heads, brushy dry weeds, structural poles, and thick grass, as well as in young trees. Titis found nonprovisioned food items only in the immediate vicinity of the grid system by looking through grasses and foliage within reach from the above-ground rails. On one occasion, a female jumped into an oak tree (approximately 2 m high) adjacent to the rail system and immediately jumped back to the grid with a leaf which she and her pairmate subsequently ate. Titis spent only one feeding interval on the ground, while 66 terrestrial feeding intervals were recorded for squirrel monkeys (<0.1 percent versus 31.7 percent, respectively).

Use of food resources. As a matter of fact, although their feeding activities were more spatially restricted, titis fed significantly more often than did squirrel monkeys (individual $\overline{X}$ intervals feeding per trial were 22.1, *Callicebus*; 5.2, *Saimiri*; $F = 14.8$; $df = 1, 8$; $P < 0.005$; mean number of feeding bouts per trial was 10.0, *Callicebus*; 4.4, *Saimiri*; $F = 8.8$; $df = 1, 8$; $P < 0.025$; see Fig. 7-22), and their bout durations were longer (individual $\overline{X}$ bout duration was 20.5 seconds, *Callicebus* versus 5.4, *Saimiri*; $F = 6.5$; $df = 1, 8$; $P < 0.05$; see Fig. 7-22). Consequently, total time spent feeding was significantly longer for titis (mean of 8.9 percent of possible time for *Callicebus*, versus 2.2 percent of possible time for *Saimiri*; $F = 12.7$; $df = 1, 8$; $P < 0.01$). Titis also consumed significantly more of the provisioned food items (pair $\overline{X}$ of 30 percent for *Callicebus* versus 13 percent for *Saimiri*; Mann-Whitney U; $P < 0.01$).

Temporal aspects of feeding. Feeding activity declined throughout the hour in both species (significant six-minute block effect for absolute frequency, interval frequency, bout duration, and proportion of time spent feeding; all $P < 0.001$; see Fig. 7-22). The decline was greater in titis (significant species x block interactions in the same measures), probably reflecting the depletion of favored resources. The lesser decline for squirrel monkeys probably reflected the transition from initial feeding on provisioned foods to a relatively con-

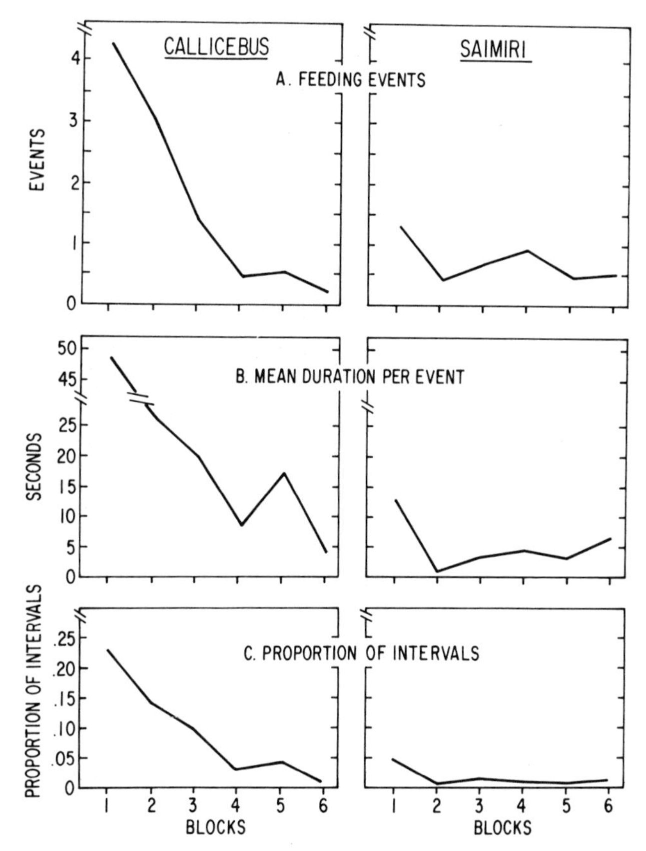

Fig. 7-22. Measures of feeding activity (subject means) per ten-minute block. (a) Feeding events taken from event-recorder data. Each event was ⩾ one second and occurred after ⩾ three seconds during which no feeding was scored. (b) Mean duration per bout calculated from event-recorder data. (c) Proportion of fifteen-second intervals on which feeding was scored.

stant, although low, rate of feeding on nonprovisioned foods outside the grid area. For example, the titis' mean bout duration declined more than tenfold over the hour, from over 47 seconds in the first ten minutes to 4.1 seconds in the last ten minutes. In contrast, the squirrel monkeys' mean duration declined about twofold, from 13.0 seconds in the first ten minutes to 7.0 in the last ten minutes. Differences in the decline of absolute frequency of feeding were comparable to those for bout duration. Mean frequency of individual feeding events per trial in titis dropped from 4.30 to 0.23 in the first to last ten minutes, respectively; comparable scores for squirrel monkeys were 1.33 and 0.53 (see Fig. 7-22).

Social aspects of feeding. The influence of social orientation was a very obvious source of species differences in feeding behavior. For example, titis fed while in contact in 64 percent of the trials and 18 percent of the intervals in which feeding occurred. Squirrel monkeys were never scored as feeding while in contact. Mean interanimal distance during feeding was much greater in squirrel monkeys than in titis, averaging 20.0 m and 5.2 m, respectively (see Fig. 7-23). Pair correlations of proportion of time spent feeding (based on twenty-four ten-minute blocks per pair) were substantially higher in *Callicebus* (*Callicebus* $\bar{X}$ pair $r_{xy} = +0.70$ versus +0.56, *Saimiri*, $P < 0.056$, Mann-Whitney U), as were intrapair correlations of frequency of feeding bouts (*Callicebus* $\bar{X}$ pair $r_{xy} = +0.57$ versus +0.38, *Saimiri*; n.s., Mann-Whitney U).

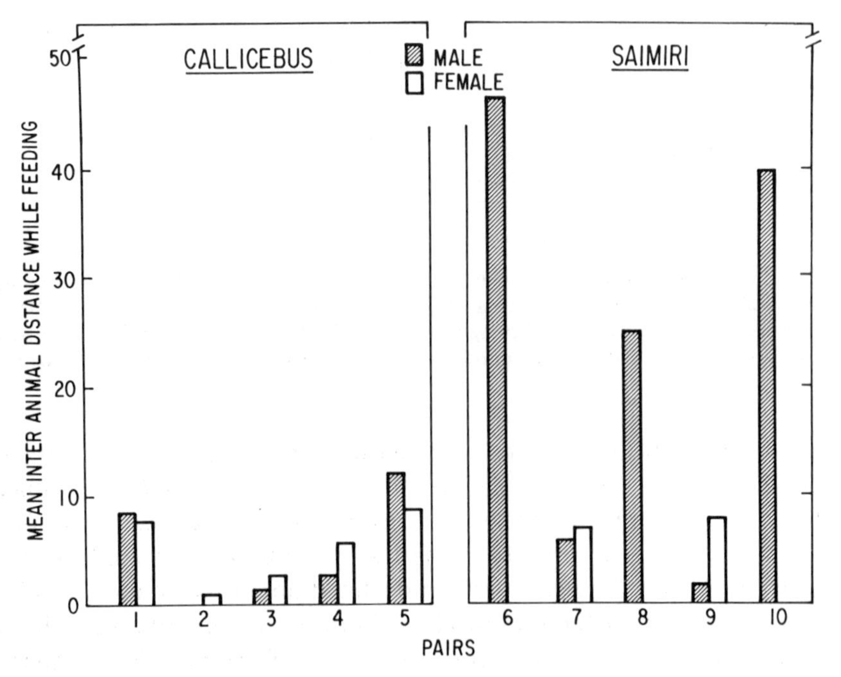

Fig. 7-23. Average interanimal distance on feeding intervals plotted by subject. Data from *Saimiri* females in pairs 6, 8, and 10 are excluded: each had ⩽ 3 intervals feeding scored over four hours.

Vocalizations and Displays. The species differed markedly in vocal and display behavior. Squirrel monkeys frequently gave high-pitched, extended "seee" calls (see Thorington, 1968; Winter et al., 1966) for several minutes upon release. Some females continued to give these well into the hour. Occasionally, separated pairmates would vocalize in sequence, but approach to a vocalizing monkey by the pairmate was not evident here, as is reported for monkeys living in seminatural conditions (Baldwin, 1968, 1971). Observers could not hear answering vocalizations from the laboratory colony several hundred meters distant, although it is possible that such sounds were audible to the subjects. Titi monkeys exhibited a greater variety of species-typical motor and vocal activities upon release than did squirrel monkeys. Loud and persistent full calls (see Mason, 1966; Moynihan, 1966) occurred frequently in all trials in one pair; they were initiated principally by the male but the female often joined in calling. When the male began calling early in the hour, the female would approach him and both would sit together calling loudly for several minutes. Her participation decreased over time. Other monkeys frequently lashed their tails, showed piloerection, rubbed their chests with their hands, and occasionally anal marked the runways, all behaviors considered to indicate a high state of arousal (Moynihan, 1966). With respect to orientation to pairmate vocalization in *Callicebus*, it is interesting to note an atypical trial where male and female were separated for a long period of time. The male made no attempt to approach or answer his active, chirping pairmate although he could see her at times and could easily hear her. She eventually sat quietly in one fence corner, and he remained quietly at his original location (hidden behind the observer platform). Thus, although close coordination of *Callicebus* pairmates is the norm, exceptions occur, even in otherwise close and compatible pairmates. Most pairmates were not visually separated for such long periods, and long-distance orientation to a calling pairmate was not an obvious mechanism for the maintenance of proximity.

Discussion

These data extend our knowledge of behavioral attributes of squirrel monkeys and titi monkeys to behavior in a large, heterogeneous novel environment. They confirm findings of major species differences in activity level obtained in more restricted laboratory settings and in a familiar natural habitat, indicating that *Saimiri* are much more active than *Callicebus* (Baldwin and Baldwin, 1971; Mason, 1966, 1968, 1971; Thorington, 1967). Differences in social orientation likewise are in agreement with those obtained in other settings, indicating that titi pairs maintain much closer spatial proximity and show greater social coordination of activities than do squirrel monkey pairs (Fragaszy and Mason, 1978; Mason, 1968, 1971, 1974; Phillips and Mason, 1976).

Differences in feeding style are similar to those observed in the living cage and in formal tests: titis are more socially attentive than are squirrel monkeys; they are slower feeders, spend more time eating a given food, and consume more of it. Squirrel monkeys are wasteful, impulsive, asocial feeders (Fragaszy, 1978). The present data show that these differences are associated with contrasting modes of resource utilization: squirrel monkeys forage for scattered nonprovisioned items in the cage interior, whereas titi monkeys feed almost exclusively on clumped, easily accessible provisioned items. Again, these findings confirm observations of frequent "foraging" on the floor of the home cage in squirrel monkeys and the less frequent retrieval of scattered food items by titi monkeys (Fragaszy, 1978) and parallel field observations of sustained traveling/foraging in *Saimiri* (Thorington, 1967) and discrete feeding bouts in particular trees, in *Callicebus* (Mason, 1966). A related difference in feeding style is frequency and mode of food transport: titis are seldom seen to carry food in the hand while moving; they rarely carry it in the mouth and then only for short distances; and they sit while eating. Squirrel monkeys frequently walk or run bipedally while holding food in their hands, and food is often eaten while on the move. This was especially noticeable at the start of the *Saimiri* trials: the monkeys grabbed banana slices from the feeder that lay along their path to the western fence and ate them while running towards the fence. Taken together, these differences support the tentative characterization of *Callicebus* as leisurely, deliberate feeders and *Saimiri* as restless, opportunistic foragers (Fragaszy, 1978).

Both species were apparently highly aroused by and engaged with the novel environments on all their trials. This was indicated by the occurrence of visual and vocal displays by titi monkeys, vocalizations in squirrel monkeys, and the high locomotor activity and low levels of social involvement and feeding in comparison to baseline home-cage levels in both species (Mason, 1974). In both species, marked adjustments in intensity and frequency of behavior occurred within each trial, suggesting a habituation and/or fatigue factor: travel, feeding, and displays all declined. Thus, both species were clearly affected by the novel setting and both reacted to it in ways suggestive of initial high arousal.

With respect to patterns of behavioral adjustments in space, both species shared sustained high attraction to figure-ground relationships (structures and edges), an element basic to visual-spatial orientation in mammals in general (see Menzel, 1966, 1969; Wilson, 1972; Walsh and Cummins, 1976). In both species, nearly all above-ground grid travel occurred along the perimeter, and nearly all nongrid travel occurred along the fence line. In both species, use of space varied as a function of exposure, suggesting familiarity was an important contributing element in use of space (Menzel, 1969; Shillito, 1963; Welker, 1957, 1961; Hughes, 1968).

However, the detailed patterns of approach and orientation to figure-ground stimuli were quite distinct in each species. Upon release, squirrel monkeys moved immediately toward the western fence and fence corner, usually pausing briefly near the observer's platform, the only large solid structure in the enclosure. Titis moved toward a feeder, often a corner feeder, on the above-ground grid system. That titis would occasionally transport food from one feeder to another before eating it suggested that feeders were attractive structures independent of the immediate value of their contents. Furthermore, viewing social spacing as a special category of orientation to objects in the environment, titis were clearly more attracted to the pairmate than were squirrel monkeys.

Species differences in both social spacing and individual location preferences suggested that the visual environment was evaluated differently by each species. It seems likely that decrease in object attraction as distance from the object increases is much greater in *Callicebus* than in *Saimiri.* Another indication of this is that *Callicebus* would sit and look at a novel object for a period of time before approaching it, or even without approaching it at all, whereas *Saimiri* approached a novel object quickly and investigated it at short range. Decreases in object attraction as a function of distance may be due to decreases either in visual interest or in willingness to expend effort to approach the object, or, as is likely, to both factors. Further study is needed to clarify the contributions of motor and perceptual interest factors to species differences in approach behavior. In any case, declining attraction to distant stimuli may be one factor which contributed to restricted travel in *Callicebus*, a territorial species. Conversely, approach to distant stimuli may have contributed to wide-ranging travel in *Saimiri.*

Changes in spatial preference over time suggest further species differences in evaluation of the visual environment. Although use of space by titis was initially even more diverse than by squirrel monkeys, increasing exposure was associated with declining travel and more restrictive use of space in this species. Furthermore, travel was restricted to small areas of intensive use, and the same small areas were occupied for successively longer periods on successive trials. This pattern contrasted with that found in squirrel monkeys: travel diversity increased over trials while travel distance remained essentially unchanged. Consequently, time spent in familiar areas did not increase over trials, as it did in titis. The picture gained from these data suggests the establishment of small, familiar, intensely used areas by titis, and conversely, the continued "exploratory" travel into unfamiliar areas by squirrel monkeys. Patterns of use of space in the novel environment thus agree with the development of the species-typical spatial patterns seen in the wild: territoriality in *Callicebus*, and ranging in *Saimiri.*

The degree of attraction to familiar visual stimuli may be one source of species differences in patterns of use of space. These data suggest that titi monkeys are probably more attentive to, or attracted to, familiar visual stimuli than are squirrel monkeys. Territoriality in *Callicebus* thus may reflect attraction to the familiar home area as well as relative disinterest in distant stimuli. Attraction to the nearby social pairmate is, of course, also likely to be affected by that monkey's familiarity, as well as by its proximity. In contrast, greater interest in novel than in familiar visual stimuli by *Saimiri* complements interest in distant features of the environment, enhancing motivation to travel and contributing to the dissolution of large groups into smaller foraging units during the day (Thorington, 1967).

On the other hand, several lines of evidence suggest that the high degree of social disinterest evident in *Saimiri* pairs, especially on the part of females, in the novel environment may have been an artifact of prior pair-housing. Attraction to a familiar male has been observed to be lower in pair-housed than group-housed *Saimiri* females (Vaitl et al., 1978). Furthermore, the presence of a male pairmate seems to inhibit female access to objects of interest in the home cage in pair-housed monkeys, but not in group-housed monkeys (Fragaszy and Mason, 1978). Thus, females in the novel environment may not only have been neutral towards their male pairmates; they may have avoided them if the males were actively engaged in environmental investigation.

The more active role taken by squirrel monkey males in following, as suggested in this study, is in agreement with previous field observations of male approach to female subgroups (Baldwin, 1968, 1971) and with laboratory data that suggest that males initiate more between-sex interactions than do females (Coe and Rosenblum, 1974; Fragaszy and Mason, 1978; Strayer et al., 1975). Even so, orientation to the familiar social partner played essentially no role in individual behavioral organization in either sex.

Social orientation data suggest that a more active role is taken by titi females than by squirrel monkey females. These data are in accord with evidence from previous studies indicating that female titis approach and follow a familiar mate more frequently than do males (Mason, 1971; Robinson, 1977). The higher proportion of female following could perhaps be interpreted as lower female interest in the nonsocial environment, but this is not supported by feeding data. Females spent more time feeding than males, particularly early in the hour when locomotor activity and following were also highest. Apparently, maintaining social proximity (requiring active social orientation) did not interfere with attention to nearby nonsocial features of the environment (feeder boxes) or with feeding activity.

Robinson (1977) has suggested that titi males are more "nervous" than females, as indicated, for example, by more frequent withdrawal from an unfamiliar human observer. Data reported here support this view. Males moved more frequently and in longer bouts than females; yet males entered fewer locations than females, especially in early trials. Most "exploratory" travel forays away from the grid area were initiated by females. Lastly, the only episode of leaping into a tree to get a food item was achieved by a female. In general, females' expressions of interest in the nonsocial environment seemed less constrained.

Summary

Behavior of monkeys of both species in the novel environment varied in significant ways from that seen in the living cage. These departures, including the temporal pattern of behavioral change, reflected the influence of the same psychological parameters in each species: tonic arousal over trials, fatigue and/or habituation within trials, sustained attraction to figure/ground relationships, and changes in spatial preference as a function of exposure. However, basic aspects of behavioral organization associated with the maintenance of species-typical life-styles were retained by each species: attraction to familiar and nearby visual stimuli, including the pairmate, in *Callicebus* versus attraction to distant and novel stimuli in *Saimiri*; leisurely and attentive feeding in *Callicebus* versus restless opportunistic foraging in *Saimiri*; comparatively low levels of travel in *Callicebus* versus high levels in *Saimiri*. These major dimensions of species differences in behavior have now been observed in a variety of settings, indicating that they do indeed reflect fundamental species-typical modes of perceiving and acting upon the environment. The same environment yields different information and provides different behavioral opportunities to individuals of the two species. Thus, the relationship between organism and environment, or the niche, occupied by each species may be viewed as the product of individual modes of perceiving and behaving in that environment. The organism-environment relationship is constrained by environmental parameters, to be sure, but it is also structured according to internal properties of the behaving organism.

REFERENCES

Baldwin, J. D. The social behavior of adult male squirrel monkeys in a seminatural environment. *Folia Primatol.* **9**: 281–314 (1968).

Baldwin, J. D. The social organization of a semifree-ranging troop of squirrel monkeys (*Saimiri sciureus*). *Folia Primatol.* **14**: 23–50 (1971).

Baldwin, J. D. and Baldwin, J. I. Squirrel monkeys (*Saimiri*) in natural habitats in Panama, Columbia, Brazil, and Peru. *Primates* **12**: 45–61 (1971).

Coe, C. L. and Rosenblum, L. A. Sexual segregation and its ontogeny in squirrel monkey social structure. *J. Hum. Evol.* **3**: 551–561 (1974).

Cubicciotti, D., III and Mason, W. A. Comparative studies of social behavior in *Callicebus* and *Saimiri*: Male-female emotional attachments. *Behav. Biol.* **16**: 185–197 (1975).

Erickson, G. E. Brachiation in New World monkeys and Anthropoid apes. *Symp. Zool. Soc. London* **10**: 135–164 (1963).

Fragaszy, D. M. Contrasts in feeding behavior in squirrel and titi monkeys. In D. J. Chivers and J. Herbert (Eds.) Recent Advances in Primatology Vol. 1. Behaviour. London: Academic Press, 1978, pp. 363–367.

Fragaszy, D. M. and Mason, W. A. Response to novelty in *Saimiri* and *Callicebus*: Influence of social context. *Primates*, 19: 311–331 (1978).

Glickman, S. E. and Sroges, R. W. Curiosity in zoo animals. *Behaviour* **26**: 151–188 (1966).

Hughes, R. N. Behavior of male and female rats with free choice of two environments differing in novelty. *Anim. Behav.* **16**: 92–96 (1968).

Mason, W. A. Social organization of the South American monkey, *Callicebus moloch*: A preliminary report. *Tulane Studies in Zool.* **13**: 23–28 (1966).

Mason, W. A. Use of space by *Callicebus* groups. In P. C. Jay (Ed.) *Primates: Studies in Adaptation and Variability.* New York: Holt, Rinehart and Winston, 1968, pp. 200–216.

Mason, W. A. Field and laboratory studies of social organization of *Saimiri* and *Callicebus.* In L. A. Rosenblum (Ed.) *Primate Behavior.* New York: Academic Press, 1971, pp. 107–137.

Mason, W. A. Comparative studies of social behavior in *Callicebus* and *Saimiri*: Strength and specificity of attraction between male-female cagemates. *Folia Primatol.* **23**: 113–123 (1974).

Menzel, E., Jr. Responsiveness to objects in free-ranging Japanese Monkeys. *Behaviour* **26**: 130–150 (1966).

Menzel, E., Jr. Naturalistic and experimental approaches to primate behavior. In E. Willems and H. Raush (Eds.) *Naturalistic Viewpoints in Psychological Research.* New York: Holt, Rinehart and Winston, 1969, pp. 78–121.

Moynihan, M. Communication in the titi monkey, *Callicebus. J. Zool. Lond.* **150**: 77–127 (1966).

Phillips, M. and Mason, W. A. Comparative studies of social behavior in *Callicebus* and *Saimiri*: Social looking in male-female pairs. *Bul. Psychonom. Soc.* **7**: 55–56 (1976).

Robinson, J. G. Vocal regulation of spacing in the titi monkey (*Callicebus moloch*). Unpublished Ph.D. dissertation, University of North Carolina, Chapel Hill, 1977.

Shillito, E. E. Exploratory behavior in the short-tailed vole. *Behaviour* **21**: 145–154 (1963).

Strayer, F. F., Taylor, M., and Yanciw, P. Group composition effects on social behavior of captive squirrel monkeys (*Saimiri sciureus*). *Primates* **16**: 253–260 (1975).

Thorington, R. W., Jr. Feeding and activity of *Cebus* and *Saimiri* in a Colombian forest. In D. Starck, R. Schneider, and H. J. Kuhn (Eds.) *Progress in Primatology*. Stuttgart: Gustav Fischer, 1967, pp. 180–184.

Thorington, R. W., Jr. Observations of squirrel monkeys in a Colombian forest. In L. A. Rosenblum and R. W. Cooper (Eds.) *The Squirrel Monkey*. New York: Academic Press, 1968, pp. 69–85.

Vaitl, E., Mason, W. A., Taub, D., and Anderson, C. Contrasting effects of living in heterosexual pairs and mixed groups on the structure of social attraction in squirrel monkeys (*Saimiri*). *Animal Behavior*, 26: 358–367 (1978).

Walsh, R. N. and Cummins, R. A. The open-field test: A critical review. *Psychol. Bul.* **83**: 482–504 (1976).

Welker, W. I. "Free" versus "Forced" exploration of a novel situation by rats. *Psychol. Rep.* **3**: 95–108 (1957).

Welker, W. I. Exploratory and play behavior in animals. In D. W. Fiske and S. R. Maddi (Eds.) *Functions of Varied Experience*. Homewood, Ill.: Dorsey Press, 1961, pp. 175–226.

Wilson, C. C. Spatial factors and the behavior of non-human primates. *Folia Primatol.* **18**: 256–275 (1972).

Winter, P., Ploog, D., and Latta, J. Vocal repertoire of the squirrel monkey (*Saimiri sciureus*), its analysis and significance. *Exp. Brain Res.* **1**: 359–384 (1966).

8
Environmental Enrichment and Behavioral Engineering for Captive Primates

Hal Markowitz

Oregon Zoological Research Center
Washington Park Zoo
Portland, Ore.

INTRODUCTION

> What is special about primates? Human interest in animals usually has an economic basis, but this group is ecologically rather unimportant, with the gigantic exception of our own species. There is a small biomass group of nonhuman primates, and they are not very numerous except in a few forested areas, nor are the members of the group very diverse in their adaptations to their environments. And yet we are so interested in them that several species are in danger of extinction from overtrapping.
>
> T. Rowell, 1972

Human-oriented research models often demand primates (because we too are primates). But far surpassing this plausible need, there has been a consistent preoccupation with, and frequent exploitation of, other primates by man. Some people have undoubtedly seen monkeys and apes as much greater potential threats than other species because of their similarity to us (e.g., the popularity of *Planet of the Apes*). In happier cases, primates are accepted as likely "comrades" because our physical similarities allow us a feeling of rapport difficult to approach with other animals. The proliferation of "language" studies with apes (e.g., Fouts, 1974; Gardner and Gardner, 1975; Premack, 1971; Rumbaugh and Gill, 1976) lends testimony to this idea. Unfortunately, traditional husbandry protocol has almost always been designed for the comfort of the researcher or the exhibitor rather than the focal species. This generates a number of serious problems from a humane standpoint and from the perspective of research and education involving nonhuman primates.

First, there is the increasing awareness that animals besides ourselves have sensitivities which demand that thoughtful human beings make careful well-informed decisions about removal from natural habitats. A corollary of this, for students of behavior and ecology, is the recognition of considerable contradictions between field and laboratory studies of the same species with respect to behaviors which are *nominally* the same. Indeed, for some field workers, there is a genuine concern that laboratory studies may have little relevance for "real life" situations. Laboratory researchers are accustomed to obtaining vernier precision with their equipment and having constant access to their animal subjects. Consequently, they are wary of field data which inevitably involve some discontinuities in observation and frequently require much extrapolation from direct observation to develop coherent models.

Many writers have convincingly suggested that a marriage of field and laboratory techniques is required to address many of the most significant questions about animal behavior (e.g., Kuo, 1967; Lehrman, 1964). One suggestion is that principles may be identified in laboratory environments and their applicability in "nature" can then be sought by field observation. For some problems, the reverse procedure may be appropriate; e.g., an apparent species capability suggested from field data may receive more fine-grain analysis with laboratory subjects. The majority of this chapter will suggest intermediate approaches in which larger than usual captive environments may be engineered to provide opportunities for species-typical behaviors (Markowitz, 1975ab; Markowitz and Woodworth, 1978).

The zoo is admittedly an artificial setting, but with proper planning, it may serve as a naturalistic intermediate ground between research in the field and the laboratory. For more than five years, the research program with primates at the Washington Park Zoo in Portland has emphasized enriching behavioral opportunities. The primary focus has been on allowing zoo residents some chance to determine their own daily schedules and providing them some control over their own environments. This chapter will summarize some of the results of work with three species: white-handed gibbons (*Hylobates lar*), diana monkeys (*Cercopithecus diana*), and mandrills

(*Papio sphinx*). Each of these examples is drawn from an environment constructed of concrete and cyclone fencing, and the solutions engineered for these animals must be considered in this context (Markowitz, 1975c). The last section of the chapter will describe the newest design plans for behavioral engineering in an emerging zoo in Hilo, Hawaii. There it will be possible to combine naturalistic exhibit techniques with equipment which provides the animals exercise and a responsive environment (Markowitz, 1977).

THE GIBBON PROJECT

Our first major behavioral engineering project chose white-handed gibbons as subjects because, in many ways, they represented the most grotesque contrast between feeding in nature and in the zoo. These beautiful apes, so specialized for brachiation and leaping, may be observed in the wild running active patterns through the trees and eating fruit which they have collected. The traditional zoo-feeding protocol (in which food was thrown through a slot in the door once a day and left on the ground for arboreal primates) certainly allowed much room for improvement. Without a large budget, and with no authorization to dynamite the cage and start from scratch, we decided to do what we could to accomplish a few fundamental goals.

First, we wanted to insure that the gibbons could eat when they wished to do so, throughout the waking day. It was also planned that food should be delivered somewhere above ground level and that some healthful exercise should be attendant to its collection. Finally, we hoped someday to allow visitors to feel some sense of contact with the animals. We wanted to accomplish this in a manner which would garner more respect for the gibbons than was represented by the usual face-making and food-throwing attempts at interaction by the public. Although these were our primary goals, there were a lot of secondary benefits which we anticipated: a long-term opportunity to record seasonal and other fluctuations in the behavior of these apes, an active full-time learning laboratory for our students, and a test case from which we could project the utility of behavioral engineering approaches to improving captive environments (Markowitz et al., 1978).

Although I heard my friends in the zoo talking about the need for "natural" exhibits, a careful examination showed that this meant natural looking. Fiberglas trees and backdrops were being promoted to satisfy the public that the animals were living a natural existence. We began what has become a long (and admittedly redundant) education program to stress the need for attention to behavioral opportunities and to emphasize that naturalistic *behavior* will not magically emerge unless we change husbandry as well as appearance of zoo enclosures. Providing an artificial tree, or even a real one, for an animal which typically feeds in trees will do little to guarantee natural behavior if we continue to throw their food upon the ground once a day.

After six years, the equipment described here frankly seems quite impoverished for an animal with the gibbons' extraordinary physical capabilities. We hope to do much more naturalistic and flexible engineering with arboreal apes in the future. The Hilo gibbon exhibit described at the end of this chapter is the first such attempt. But, it is important that this be reviewed in context. The choice was *not* leaving these animals in Southeast Asia versus providing some artificial regimen for exercise. Instead, the alternatives were leaving a cage in which the gibbon had no control over its life or providing changes (without major remodeling) to insure that the environment was somewhat more responsive. Two apparatus arrays were installed approximately 8 m apart and $4\frac{1}{2}$ m high on the back wall of the cage. Each of these panels included a large stimulus globe and lever. The right-hand apparatus also had a food dispenser. The gibbon colony at that time consisted of two mature males, one female, and an infant male.

We were faced with some unusual training requirements because we did not want to separate the animals. It seemed desirable to develop a situation in which the *gibbons* could establish the rules about cooperation, sharing, or individual food earning. With careful observation, we were able to find opportunities to train each of the gibbons to turn off the light at the right-hand station in order to obtain pieces of their regular diet (apples, bananas, onions, carrots, oranges, and monkey chow). It was only a few days before all three full-grown cage residents were successfully earning food and eating high up in the cage.

Next, we proceeded to the final portion of training in which we wished to encourage more active brachiation and leaping. To initiate movement between the two stations, we began by delivering food for movements away from the payoff station toward the left-hand apparatus. Despite some dire predictions from colleagues, we found the gibbons so capable that in about a month, each was able to accomplish the entire sequence (see Fig. 8-1). Our gibbons were active, feeding up high, and delivering us reams of interesting data about the "social conventions" which they developed.

A few examples may provide some flavor of the richness of this data and suggest similarity to observations for many primate species in the wild. One of the adult males, Harvey, had always been the most agile and adept swinger in this group. He became so proficient at earning his own food that he was able to complete the entire se-

Fig. 8-1. Apparatus to encourage activity and allow gibbons to schedule food earning. Harvey responds at the left-hand station.

quence in about $1\frac{1}{2}$ seconds. Shortly, his mother and his brother began to exploit him by going to the payoff station and waiting for Harvey to make the big excursion. Harvey's response to this treatment surprised us no end . . . he would often work for many trials allowing his mother free access to food, but after a few weeks, he no longer provided for Kahlil. Instead, Harvey would back away until Kahlil had left the payoff station and moved far enough across the cage to allow an active "race" for food. We often saw pieces of fruit broken between the gibbons and shared after these contests. Thousands of hours of interesting observations (attempting to differentiate sharing from stealing, etc.) were logged. The important point was that the gibbons were developing their own solutions to a life which allowed them to feed ad lib in return for a little exercise.

Quite by accident, we were provided an opportunity to evaluate the gibbons' interest in the responsiveness of the apparatus as opposed to its simple functional utility in delivering them food. This opportunity came when we decided to change our controls to a new set of solid-state equipment. During the period of transition, we supplied bite-size pieces of food up high ad lib to the gibbons. Frequently, often with free food in their hands, they attempted to get the lights and levers to respond to them, even occasionally hugging the globe while pulling the lever next to it. For the reader who may find this miraculous and totally unexpected, we would suggest an analogy to the frequently replicated work of Neuringer (1969) in the laboratory. It is now fairly well accepted that many animals, given a choice between freely available identical food and working in simple ways for the food, opt for the latter. From an anthromorphic standpoint, it seemed clear to us that the gibbons missed the opportunity to actively change their own environments and to "produce" their own food.

Several months later, the public was allowed an opportunity to participate in the feeding. With some reluctance, I acquiesced to the idea that we should allow small contributions to initiate the availability of the food. This was accomplished by the introduction of a carefully constructed graphic:

Research Contribution:

Ten cents will start a trial when the light on this box is lit. The counter shows the total number of pieces of food earned by the gibbons today.

> Animals are not machines and the gibbons may not choose to respond when the light is turned on. All money collected here will be used to develop more activities for our animals.

A coin box in the visitor area allowed the initiation of a trial which previously had always been accomplished by a timer. However, this in no way diminished the gibbons' opportunity to feed since the same two-minute intertrial interval applied if the public chose not to contribute. Thus zoo visitors provided a free random interval generator, the gibbons began to pay some attention to visitor behavior, and they contributed in unexpected quantity to the development of similar activities for other animals. During the first year, more than $3,000 in dimes was contributed. The only complaint from a zoo visitor came when the apparatus was being serviced and he was unable to put a dime in the machine after returning to the zoo especially for this purpose (Markowitz and Woodworth, 1978).

This first venture provided us considerable experience and a wealth of questions, many of which are still unanswered more than five years later. We had developed a feeding system which was highly economical, convenient for use in the zoo, and easily applied to a wide variety of species (Markowitz, 1973). The problem of excess food lying around and decaying on the floor had been reduced to a minimum. And, our veterinarian and visiting health scientists often provided enthusiastic endorsement for the activity and natural-like feeding which this highly artificial apparatus allowed. Most prominent among our questions as we proceeded to engineer simple changes in a second primate exhibit were those concerning sharing and general altruistic behaviors as opposed to stealing and overt aggression. We had never seen any bloodletting between our gibbons, but we had observed a number of occasions where one animal was obviously disquieted by another's taking the food it had earned, and other occasions where the food earner seemed quite indifferent to sharing.

THE DIANA MONKEY EXHIBIT

We decided to work with diana monkeys, where we had another relatively spacious cage and a nice family group comprised of an eight-year-old male (Rocky), his sixteen-year-old mate (Beulah), and their adolescent male (Butch) and infant male (Kid) offspring. During the four years which we have been working with these animals, they have had an offspring annually, illustrating an important principle which we have not sufficiently emphasized above: the philosophy of behavioral engineering which we have followed has always provided ample opportunity for the animals to ignore the apparatus and engage in other activities. In a small percentage of the waking day, plenty of food to meet all nutritional requirements can be obtained. Consequently, although we make detailed daily observations and food counts, it has seldom been necessary to do any supplemental feeding except at the times of parturition, special health examinations, etc.

Apparatus for the diana monkeys was quite similar to that for the gibbons in some aspects although in this case, the first station was a chain and a globe suspended from the top of the cage at a height of about 6 m. The second station was mounted approximately 4 m high on the back wall at the right-hand side of the cage. It included two stimulus globes, a lever, and a token dispenser. These two components were trained in much the same way as we had trained the gibbons to earn their food by movement between two stations. However, this apparatus did not directly provide food, but instead provided large plastic tokens which could be exchanged for food whenever the dianas desired. This exchange was accomplished by depositing the tokens into a slot in a food dispenser specially designed for installation in the cage door at an elevation of about 6 m (see Fig. 8-2).

The dianas were first trained to deposit tokens and this turned out to be a very laborious procedure. We tried balancing the tokens in the slot for the animals, occasionally dropping them for them, and putting our hand inside the cage and allowing opportunities for imitation. Finally we discovered, by trial and error, that depositing tokens (even into a rather wide slot) was apparently a difficult task for this species. We redesigned the apparatus with a V-shaped slot which helped to orient the coins for the monkeys, and in one day, all but the female and the infant were readily depositing tokens. Despite many shaping sessions, one of the significant anomalies of the behavior in this cage, throughout what is now several years of work, has been the females' inability to learn to deposit tokens. We finally decided to proceed to the next stage of training although Beulah had not acquired this response. As will be seen from the brief behavioral

Fig. 8-2. Diana monkeys engage in a token economy. Adult male deposits token in food vendor.

descriptions below, this did not diminish the richness of Beulah's role in cage dynamics.

The dianas next received forty-three half-hour training sessions at the token delivery station and became quite adept in obtaining and spending tokens in this period. We also began to see the emergence of surprising behaviors which have persisted through the years. For example, Rocky initially dominated the scene until he was satiated. During these periods, he would not tolerate the youngsters' attempts to share any of the food he had earned with tokens. If they sat to accompany him at the payoff station, he pushed them away, deposited the token, and took the food himself. On the other hand, his mate was sometimes allowed to sit next to him while he deposited two tokens, letting her have one of the pieces of food. Soon the juvenile also became adept at token earning and exchange, but he had his mother to contend with. She would often take the food that he earned and Butch was left with no alternative but to go earn another token.

Finally, the last station at the top of the cage was activated and it took little time for the animals to acquire this additional response component because the youngsters would regularly swing on the chain when the lights went on. With the full sequence now available to the animals from morning to dusk, we were treated to behavioral displays unparalleled in other diana monkey exhibits (Chasan, 1974). Until he became more comfortable with the constant availability of tokens and food, Rocky would often horde tokens. He was once seen carrying four with one foot and three between his teeth. This behavior did not persist for very long, because all of these active monkeys became so good at earning tokens that they would earn them "for the fun of it" and let them lay around to be spent by others later. This was especially true in later parts of the waking day when the animals were more satiated.

During the earlier hours, Beulah stepped up her acquisitiveness, but not by learning how to spend tokens. She would often go so far as to pat one of the youngsters on the behind, encouraging it to run down and spend a token, and then would "rip off" their food. Occasionally someone was distressed by this behavior, thinking that we might have taught this monkey to be a thief. An effective response was to visit an adjacent cage where a related species of monkeys was feeding from piles of food on the floor. Here, youngsters were typically pushed aside when they reached for the more choice morsels, and the food was clearly dominated by the older animals until they were satiated.

Another interesting aspect of the dianas' behavior was the difference in the flexibility shown in learning how to earn tokens. For many months, Rocky always approached the chain in exactly the same way. No matter where he began in the cage, he would move to the right-hand side, walk along the same series of bars, *always* turn right, and finally pull the chain. Then he would turn around, retrace his steps, turning left this time, and take a slow methodical path to the second station. Butch showed much greater flexibility, quickly learning that there were all sorts of shortcuts, such as leaping at the chain, curling his legs around the bar beneath it and swinging in one loop to the second station. I felt very fortunate that during one of the two-hour daily observation periods, I was present to watch an interesting change in Rocky's methods. He sat at the point where some of the bars in the cage converged, about 2 m from the chain, and watched while Butch went through some gymnastics earning tokens with his "short loop" method. After several tokens were earned in this fashion, Rocky finally went over to the chain, waited for the light to go on, pulled it himself, and for the first time, took the shortcut. I know that I am inviting criticism about excessive anthropomorphism, but I cannot help expressing to you in an anecdotal way, the look of "pride" which father gave son after this accomplishment. Although he occasionally returned to his stereotypic route, Rocky now became increasingly flexible and eventually took many different paths in token and food earning.

I have intentionally used much space in describing anecdotal outcomes of this work to try to illustrate a most important point. The incorporation of behavioral opportunities for primates allows for a wealth of observation which zoo visitors and researchers may share. Rather than diminishing behaviors other than those required for food earning, all sorts of cage dynamics are potentiated. The zoo visitor gains respect and even admiration (Chasan, 1974) for the capabilities of the monkey. The monkey remains active and healthy, and the researcher is given the opportunity to observe many species-typical behaviors which ordinarily drop out because of traditional captive husbandry. It is unlikely that a diana monkey would ever encounter stimulus lights, chains and levers, and token dispensers in West Africa. But, it is even more unlikely that a wild monkey would live in a situation where its food was provided without activity, or where its own responses made no differences to its welfare.

THE MANDRILL GAME

Perhaps the most unnatural of stopgap procedures to overcome an unresponsive environment is the speed game which we have installed for our mandrills. All of us can think of many wonderful ways to design new habitats for this large, mostly terrestrial primate. In the meantime, while they must live in a cage, we thought it would be nice for them to have a game for entertainment. As a fringe benefit, the public is entertained, spending more than twice as much time in front of this cage as other comparable exhibits (Zawel and Tavakoli, 1977).

A major part of this game was taken from an earlier set of work by Bill Myers (1977) with a baboon. Our procedure differs in that the rules clearly make the game available to the mandrills at their option and guarantee a game even when no human beings are present. Exclusive of its controls and recording equipment, the game has three major components: two essentially identical consoles for animal and visitor, and a scoreboard (see Fig. 8-3). Although the apparatus is installed in the home cage where any of the mandrills theoretically have access to it, functionally the male never allows anyone else to play. A game proceeds as follows: a large circular panel lights and remains lighted until the mandrill decides he wishes to play. When the mandrill presses the panel, it goes off and a similar panel lights for the public. Zoo visitors have only fifteen seconds in which to ask to compete with the mandrill by depositing a dime in a slot in this panel. If no visitor chooses to play, the computer automatically generates a contest. In either case, the results are illustrated on a large scoreboard. The contest consists of lights going on at identical times on the animal and the visitor panels. The light appears in one of three random locations and premature responses automatically lose. The first competitor to touch the lighted panel wins. The contest champion is the one who first accumulates three wins.

The mandrill's acquisition of this game was phenomenal. Within two weeks, he was playing complete, competitive contests. Within a few months, he was regularly beating the public. Now, eighteen months after its original installation, the mandrill wins more than 70 percent of the contests against visitors. In competition with the computer, he has won more than 70 percent of the games at computer speeds of 0.315 seconds. When you compare this with traditional reaction-time measures for simple single-location responses (approximately 0.250 seconds for humans, chimps, and rhesus) it is not surprising that the mandrill wins the majority of the time from the public. He likes to play so much that on most days, in the early afternoon, we have to put a sign up saying, "We don't want him to get fat" because he has eaten all of the food that the veterinary staff feels is healthful. The mandrill's interactions with the public are interesting and suggest his competitive nature. When he occasionally loses, he may give shoulder shrugs and sometimes even a gape threat. When he wins, he simply looks disdainfully over his shoulder.

As with our other behavioral engineering, this exhibit has provided the opportunity for numerous "piggyback" studies. Wherever possible, we try to orient these studies to further evaluation of the effects of our apparatus on animal welfare. One such study focused on the changes in behavior of the mandrill's mate and the relative space usage and frequency of stereotypic behavior by these two animals (Yanofsky and Markowitz, 1978). In essence, this study showed that there were significant decreases in pacing and increases in active behavior, except for the fact that the female was no longer constantly chased from resting positions by the male.

We have developed new direct tape-transport systems for daily computer analysis of the mandrill's effectiveness with the game. This allows us to do immediate scanning for changes in behavior which may be indicative of health problems or changes in cage dynamics which have been missed during our two-hour daily ratings of behavior. One of the most important attributes of an active living situation for zoo animals is the opportunity to observe "signs" of health and husbandry problems. Since one cannot depend on verbal report of symptoms, routine information about expected daily behavior from animals provides the next best thing: an opportunity to tell when a resident is not acting "normal" (Markowitz et al., 1978; Markowitz et al., 1975).

THE PANAEWA ZOO—NATURALISTIC BEHAVIORAL DESIGN

When Jim Juvik assumed responsibility for planning the new Panaewa Zoo on the big island of Hawaii, he contacted me about the desirability of incorporating active behavioral opportunities for the future residents. As we all got to know Jim better, we found his plans very close to our own thinking about the ways zoos should evolve. There was the shared notion that zoos should not strive to have one of every possible species exhibited. Instead, a much smaller selection

(b)

(a)

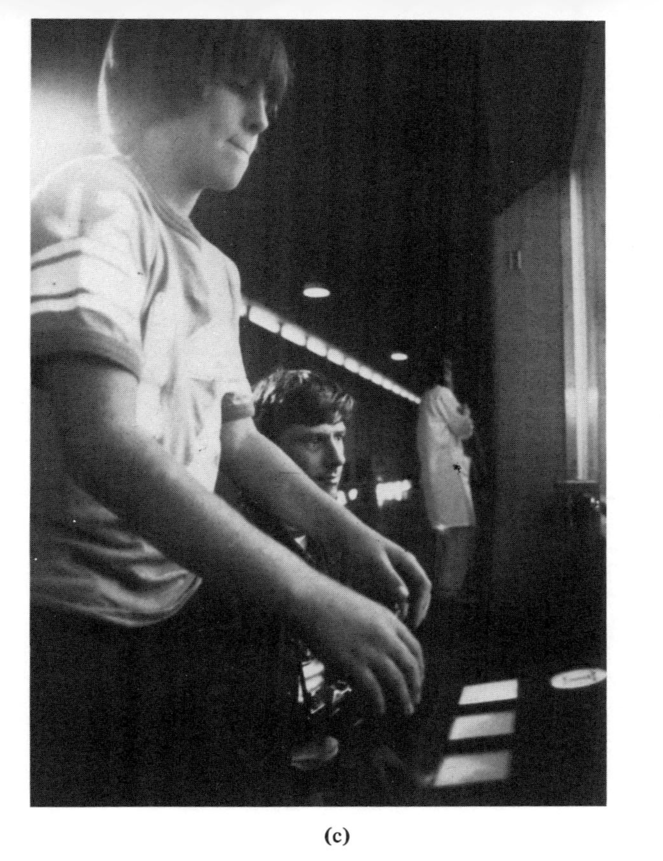

(c)

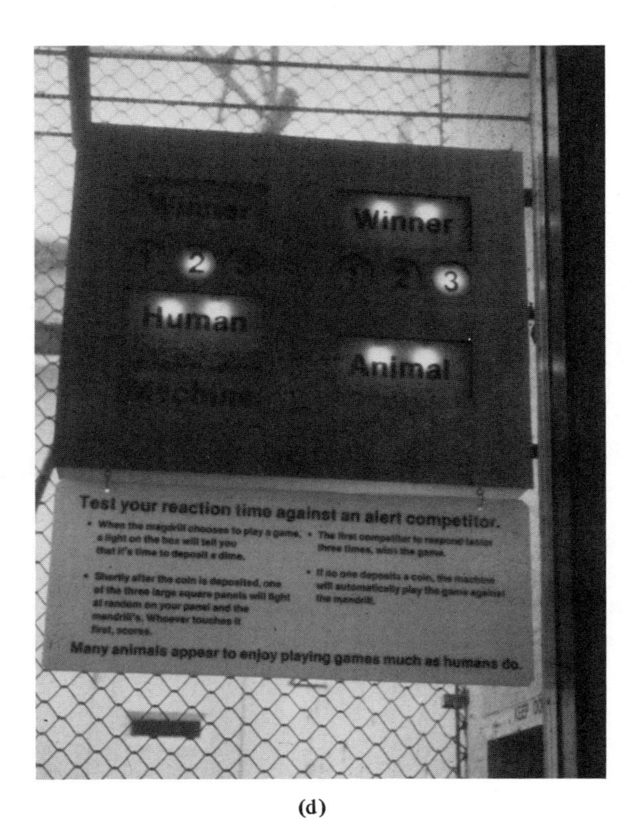

(d)

Fig. 8-3. The mandrill speed game. (a) Mandrill initiates games. (b) Mandrill ready to respond. (c) Human contestant ready to respond. (d) Outcome displayed on scoreboard.

with ample room and excellence in exhibit was to be stressed. Jim's training in climatology and the positive influences of Jack Throp, who has worked to develop much of the zoo system throughout Hawaii, insured a positive emphasis on taking advantage of the abundant natural resources found there. Instead of trying to transform the rain forest of Hilo to African plains or arctic exhibits, the first two major undertakings anticipated what Hilo has plenty of: rain.

Lush area is being transformed into a South American rain forest and a Sumatran swamp. The behavioral engineering projects which we will focus on are parts of the Sumatran swamp. This exhibit, which approaches an acre in size will be inhabited by two major species: Sumatran tigers (*Panther tigris sumatrensis*) and siamangs (*Symphalangus syndactylus*). Exhibits with mixed species always represent special husbandry problems, and these are particularly prominent when one of the animals might naturally be expected to consider the other as prey. It seemed obvious to us that unless we wanted the tigers to spend most of their time pursuing the siamangs, we had better find other ways to keep them interested and well fed. A second major problem was delivering health care for agile apes living in large open spaces. Our engineering would have to include some routine way to provide veterinary access to the siamangs. Even the lush tropical forest could not be expected to provide a perpetual food supply for both species in limited space, so we needed ways to insure the availability of healthful foods. Many of the goals of our program were dictated by these husbandry needs. All of us further believed that design conceptions should include everything that we could do to make the environment look like the animals' area of origin. The environment should also provide opportunities for use of their natural abilities. In the best initial ways we knew how, we planned to provide an educational and recreational experience for the zoogoers which would make them feel close to, and perhaps part of, the animal's life. We started with the tigers, which I will "smuggle" into this book on primates because much of the engineering was based on their sharing the same environment as the siamangs.

A bamboo grove and trees were preserved during the construction phases, and this forested area rises about 6–10 m above the level of a pond which surrounds the siamang island. We plan to concentrate the tigers' activities in this elevated portion of the exhibit and to take as much advantage as possible of the naturalness which the forest provides.

Four pieces of equipment were constructed for installation in the upper swamp. The first was designed for installation on a tree deep in the forest but observable from the public pavilion. In the wild, tigers frequently rear up and scratch on trees or other objects. This species'-typical behavior is much less frequently seen in captivity. Sometimes this is because there literally are no convenient objects to invite the behavior, and in other cases because the animals are so overfed that there is little motivation for exercise. An auditory sensor allows the tree to detect when it has been scratched. This response will be used to signal the tiger's interest in eating. As with our other exhibits, this will be a laissez-faire system in which the tiger can select the intervals when it wishes to eat. A second detection device will tell us when the tiger has moved across a small cliff which has been constructed to provide some interesting patterns in exercise. About 10 m on either side of this cliff artificial ground prey have been installed.

Designing equipment which will resist the ravages of a tiger and the corrosive elements of a rain forest took considerable time and effort. Ron Fial and I worked on literally hundreds of different ideas before settling on these prototypes. With the help of Nick Lee and Mike McBride, we were finally able to construct equipment which we believe will do the job. A rabbit and a squirrel model were first sculpted in clay and then cast in a tough resin. These "animals," which were approximately life-size, had to be made with very special contours. We wanted them to give the appearance of an animal in motion. At the same time, we did not want any sharp parts or crevices where a tiger might injure itself in the "capture." Steel bars were cast into the centers of the rabbit and squirrel allowing for connection to mechanical drive components. Two artificial hills were built with rebar and heavy-gauge wire mesh. These hills, approximately 1.33 m in diameter and 0.5 m high, were covered with the rich soil of the Hilo swamp terrain mixed into a concrete composition for durability. Underneath these hills were the homes of the rabbit and squirrel, securely anchored in the ground and provided with drain fields. When signaled, these portions of the exhibit give the appearance of an animal running across the top of a little hill. The tiger can "capture" these ground prey by pouncing on them or swatting at them during any of the time that they are above the surface of the ground. This capture is immediately detected by our computer and the apparatus is programmed to "save the life" of the rabbit or the squirrel for another capture.

It may be easiest to shift now to look at things from the perspective of the zoo visitor in order to describe the total program. When visitors enter the pavilion overlooking the tiger area of the swamp they will find a set of graphics which include a TV presentation. Ordinarily, the TV segment will include some general ecology lessons about the Sumatran swamp, some information about the feeding habits of the tiger, and a description of some species'-typical behaviors to watch. When the tiger scratches on the tree indicating that it feels like eating, the TV narrative will change and tell the public of the tiger's interest. It will also tell visitors that they may participate, if they choose, by selecting the order in which the rabbit, the cliff, and the squirrel become responsive activity areas. If zoo visitors do not initiate these choices within a brief time, the computer will select a random order.

Thus, with no predictability in terms of sequence, the rabbit or the squirrel may begin to run across their hills, or the cliff may "beckon" the tiger with an auditory cue. Each of these portions of the exhibit will remain active until the tiger responds to them. Successful completion of the sequence leads to the automatic delivery of fresh meat from a refrigerated device installed underneath the visitor pavilion. The appearance to the visitor is of the tiger racing toward them after capturing some artificial prey. When it has disappeared from their line of sight, the tiger can take its earned food and eat it wherever it pleases.

One final feature of the apparatus is that it is designed to allow the public to participate in the training process. During the period when the tiger is learning to capture the squirrel or the rabbit or to move across the cliff, the computer provides the flexibility to allow public interaction. One day, we can invite the public to remotely participate by initiating the rabbit whenever they choose, another day the squirrel, etc. This precludes our worrying about intensive shaping schedules or any special deprivation to teach the tiger how to earn food. We can progress through each activity at the tiger's leisure and allow the public to feel that they have a role in helping the tiger to learn. All of these activities have been built into portions of the swamp which minimize any possibility that the siamangs will become involved as active prey. The fresh meat used for reinforcement is not a preferred food for siamangs and the activities are intentionally prominent and noisy enough to make young apes wary of interaction with them.

Switching now to describe the apparatus for the siamangs, we move to the island portion of the swamp. This island will have a large pole and tree structure with a half dozen widely spaced vines hanging from it. Connecting the island structure to the siamangs' hut is a cable approximately 26 m long. This stainless cable will be used to support the siamangs' weight and to provide an avenue for the "intelligent" vines to communicate with the control apparatus on shore. The cable will be wrapped with hemp and other strong materials to provide a more natural appearance and better handholds for brachiation.

Given the opportunity, siamangs like other gibbons, are wonderful leapers and swingers, and we wanted to leave the choice of behavioral patterns to them. The first stage of our control apparatus for this exhibit is very simple. It allows the zoo staff to select any number of vines from two to six. The response required from the siamangs amounts to movement between any vines of their choice until the total number is accumulated. For example, in the case where the number 5 has been selected by the staff, the siamang, in swinging and leaping around the island, must eventually swing on five different vines to earn food. The food is delivered many meters away in the hut on shore.

Locus of food delivery was one of the most difficult decisions in engineering this project. In many ways, we would have preferred the naturalistic appearance of food delivery out on the island simulating the collection of fruit in the wild. But, the concern about husbandry outweighed this preference. We knew that we wanted to be able to have regular opportunity to examine the siamang's health at close hand, to capture it for treatment when necessary, and to accomplish routine X-ray, stool examination, and blood-sampling procedures. It was decided that the best compromise was to allow the siamangs to exercise as long as they chose and accumulate all the food that they wished in their hut before returning to it. An auditory signal was provided to let the apes know each time they had earned a piece of food.

In summary, the siamangs can earn food all day long, whenever they choose, for a preselected amount of activity. Delivering the food in their home hut means that they will return there many times a day and on those occasions where examination is necessary, one needs only to close the door behind them. We believe, based upon some pilot observations, that the great frequency of food earning relative to capture means that the siamangs will be less wary of this method of restraint than they would be of nets, etc. Indeed, one of the biggest sources of problems in traditional capture of apes involves the chase which must be as frightening to the animals as it is to their potential captors. Eliminating the need for pursuit will largely resolve this problem.

One interesting empirical question which we look forward to answering is the extent to which the siamangs will become parsimonious with their responding. Since the only information available to them will be the "payoff" signal, one possibility is that their total level of activity will simply be higher as the response requirement is increased. On the other hand, it is possible that gibbons, which we have already seen are capable of so many complex learning processes, will surprise us and learn that they must maximize movement between distinct vines. One good strategy, for example, would be to always move through all of the vines before returning to any individual one since this would always lead to maximum payoff regardless of preselected number. The opportunity to look at such questions is seen as a fortuitous bonus. Our real goal was to produce the simplest apparatus which would generate opportunities for naturalistic exercise and allow the delivery of nutritious foods.

CONCLUSIONS

This chapter has illustrated a number of aspects of behavioral engineering for environmental enrichment with a small number of selected examples. These major points may be summarized as follows.

1) Rather than diminish other healthy activities, allowing the ani-

mal some control over its own environment may actually have positive "spinoff."

2) Animals with little or no deprivation are often very willing workers, especially in an environment which is relatively dormant except for these behavioral opportunities.

3) The noisy zoo environment can yield surprisingly systematic results to evaluate the animals' behaviors in situations richer than the typical laboratory.

4) A by-product of this work which has a major goal of improving the animal's lot, is a set of unparalleled opportunities for student training and active research observation.

The last example has illustrated some of our goals in designing new exhibits. Most prominent is the fact that naturalism in design of exhibits *must* include opportunities for species'-typical behaviors. These behaviors will not magically emerge by the introduction of trees or other natural, but unused, environmental proliferation. Some contingency between the animal's behavior and their obtaining desired outcomes from the environment should be provided in captivity. Only in this fashion can we really expect regular healthful activity from our animals as they mature.

REFERENCES

Chasan, D. In this zoo, visitors learn, though no more than animals. *Smithsonian*, July 1974, 22–29.

Fouts, R. Language: Origins, definitions and chimpanzees. *J. Hum. Evol.* **3**: 475–482 (1974).

Gardner, A. and Gardner, B. Early signs of language in child and chimpanzee. *Science* **187**: 752–753 (1975).

Kuo, Z. *The Dynamics of Behavior Development.* New York: Random House, 1967.

Lehrman, D. The reproductive behavior of ring doves. *Sci. Am.* **211**: 48–54 (1964).

Markowitz, H. Biological and behavioral research with captive exotic animals. Paper presented at the Meeting of the American Psychological Association, Montreal, Aug. 1973.

Markowitz, H. New methods for increasing activity in zoo animals: Some results and proposals for the future. *Centennial Symposium on Science and Research, Penrose Institute, Philadelphia Zoological Gardens.* Topeka: Hill's Division Riviana Foods, 1975a.

Markowitz, H. Analysis and control of behavior in the zoo. *Research in Zoos and Aquariums.* Washington: National Academy of Science, 1975b.

Markowitz, H. In defense of unnatural acts between consenting animals. Proceedings of the 51st Annual AAZPA Conference, Calgary, 1975c.

Markowitz, H. On natural zoos and unicorns. Keynote address presented at the meeting of the Western AAZPA, Seattle, 1977.

Markowitz, H., Schmidt, M., and Moody, A. Behavioral engineering and animal health in the zoo. *International Zoo Yearbook.* London: Zoological Society of London, 1978.

Markowitz, H., Schmidt, M., Nadal, L., and Squier, L. Do elephants ever forget? *J. Appl. Behav. Anal.* **8**: 333–335 (1975).

Markowitz, H. and Woodworth, G. Experimental analysis and control of group behavior. In Markowitz, H. and Stevens, V. (Eds.) *The Behavior of Captive Wild Animals.* Chicago: Nelson Hall, 1977.

Myers, W. Applying behavioral knowledge to the display of captive animals. In Markowitz, H. and Stevens, V. (Eds.) *The Behavior of Captive Wild Animals.* Chicago, Nelson Hall, 1977.

Neuringer, A. Animals respond for food in the presence of free food. *Science* **166**: 339–341 (1969).

Premack, D. On the assessment of language competence in the chimpanzee. In Schrier, A. M. and Stolnitz, F. (Eds.) *Behavior of Nonhuman Primates*, Vol. 4. New York: Academic Press, 1971.

Rowell, T. *Social Behaviour of Monkeys.* Baltimore: Penguin Books, 1972.

Rumbaugh, D. and Gill, T. The mastery of language-type skills by the chimpanzee (*Pan*). *Annals of the New York Academy of Sciences* **280**: 562–578 (1976).

Yanofsky, R. and Markowitz, H. Changes in general behavior of two mandrills (*Papio sphinx*) concomitant with behavioral testing in the zoo. *Psychological Record* **28**: 369–373 (1978).

Zawel, M. and Tavakoli, C. Preference testing in *Homo sapiens* to determine popularity of zoo displays. Unpublished manuscript, Reed College, 1977.

9 Great Apes in Captivity: the Good, the Bad, and the Ugly*

Terry L. Maple

Department of Psychology, Georgia Institute of Technology, and Yerkes Regional Primate Research Center, Emory University Atlanta, Ga.

INTRODUCTION

> In the milieu of captivity, the most important factor is man; apart from his immediate presence, the whole environment of the captive animal is as it were impregnated with man. Under such circumstances, the most important behavior pattern in freedom, flight from man, the enemy, becomes meaningless. . . . The main problem set the animal in the reconstruction of its subjective world to suit captive conditions is to fit man into the new set of circumstances.
>
> H. Hediger, 1950

In Hediger's pioneering volume *Wild Animals in Captivity* (1950), he makes the point that: "However paradoxical it may sound . . . the free animal does not live in freedom, neither in space nor as regards its behavior toward other animals." Accordingly, all wild animals are constrained by the natural limits of their home range, ecological niche, territory and the further constraints provided by space, time, resources, competition, and natural enemies. In considering, therefore, the plight and welfare of captive animals, Hediger's crowning achievement has been the recognition that those principles which apply to the wild apply as well to the zoo, circus, or domestic corral. To their credit, the animal trainers Marion and Keller Breland (1966)

*The author gratefully acknowledges valuable assistance from the following grants, institutions, and colleagues: NIH grants RR00165 and HD00208; McCandless Fund for Biological Research; Atlanta Zoological Park; Zoological Society of Atlanta; Kingdom's Three Animal Park; Thomas J. Watson Foundation; and G. H. Bourne, S. Clarke, L. D. Clifton, M. B. Dennon, S. Dobbs, M. P. Hoff, R. Jackson, C. Juno, R. D. Nadler, S. Puleo, J. Roberts, R. Schonwetter, K. Southworth, M. Wilson, S. F. Wilson, and E. L. Zucker. Informative conversations with Profs. R. K. Davenport, Jane Goodall, William A. Mason, and Robert Sommer are also gratefully remembered.

also recognized this fact in their attempts to train exotic species: "... we have found that you cannot understand the behavior of the animal in the laboratory unless you understand his behavior in the wild."

To acknowledge that a captive primate is still a wild primate is the first and necessary step in providing for its needs. However, we should not misunderstand Hediger's words. The natural constraints of which he wrote were not a rationalization for bars and barren walls; the solitude that is cement. Hediger was arguing for an enlightened and informed approach to animal management. To admit constraints is *not* to encourage deprivation, suffering, or isolation. Successful management requires that an institution provide adequate space, companionship, and environmental complexity so that the animal may engage in a normal array of activities and ultimately bear and rear its offspring. Captive animals should furthermore be free of psychopathological behaviors and stress diseases which are the by-product of inadequate rearing conditions.

In this chapter, I will endeavor to outline the "state of the ape" in captivity, and, in so doing, offer recommendations for their proper management. Some of the ideas here are original, others have been obtained from the literature of animal behavior and husbandry. It is my hope that, whatever the source, the ideas presented here will be of use to those who carry the special responsibility of ape-keeping. In a more general sense, these points and principles will very often apply to the management of other species, and I trust that they will be of similar utility. In preparing this contribution, I am in agreement with Van Hooff (1973) in acknowledging "a growing feeling in the last few years that the conventional methods of keeping great apes ... are not fully adequate." Indeed the present deficiencies are more prevalent than the few improvements, however spectacular.

A BRIEF HISTORY OF APE-KEEPING

While both chimpanzees and orangutans were exhibited in Europe as early as the eighteenth century, few specimens lived for very long. Indeed, the failure to observe apes for any appreciable period of time, in the wild or in captivity, led to considerable confusion over the number of extant species. A common misconception was the proposition that young female apes were "pygmy" or "dwarf" species. Furthermore, the jumbled-up taxonomy of the eighteenth, nineteenth, and early twentieth century is an obstacle to an accurate historical record, since apes were often inaccurately portrayed as representative of other species. For example, Lang (1959) argued that the first recorded exhibition of a gorilla (1855) was, in fact, that of a chimpanzee.

In the public display of chimpanzees, gorillas, and orangutans, the major obstacles to survival were their susceptibility to human disease, inadequate diets, irregularities of climate, and, undoubtedly, the effects of stress. In this century, the earliest breeding and housing successes were recorded by Madame Rosalia Abreu in Cuba (see Fig. 9-1). Madame Abreu's colony at Quinta Palatino was described by Robert Yerkes in his 1925 volume *Almost Human*:

> If, then, we were asked to sum up for the mistress of Quinta Palatino, as well as ourselves, the essentials of success in keeping and breeding the higher primates, we should emphasize the following points: freedom, or reasonably spacious quarters, fresh air and sunshine, preferably coupled with marked variations in temperature; cleanliness of surroundings as well as of the body; clean and carefully prepared food in proper variety and quantity; a sufficient and regular supply of pure water; congenial species companionship and intelligent

Fig. 9-1. Chimpanzee house at Abreu's Quinta Palatino colony (from R. M. Yerkes, 1925).

and sympathetic human companionship, which, transcending the routine care of the animal, provides for the development of interest if not friendliness; and, finally, adequate resources and opportunity both in company and in isolation for work and play.

In nearly every instance, as we will see, this advice is sage indeed. Yet, with few exceptions, the captive display of great apes has remained, at worst, a public disgrace, and, at best, a relic of less enlightened times. It is with candor, therefore, that I portray our *entire* history of ape-keeping as an example of the good, the bad, and, *all-too-often*, the ugly.

In summarizing the accomplishments of Madame Abreu, Yerkes added that: "Given these conditions of captive existence, primates originally healthful and normal should without difficulty be kept in good condition of body and mind and should naturally reproduce and successfully rear their young." We have come to judge our success in ape-keeping in exactly this fashion: successful *bearing and rearing of offspring.* The psychologist, however, may well take a dim view of the narrowness of these two objectives. Research into animal psychopathology (cf. Berkson, 1968; Erwin et al., 1973; Maple, 1977; Rogers and Davenport, 1969) has broadened our understanding of stereotyped motor acts, autoerotic and autoagonistic behaviors. Therefore, as earlier asserted in this chapter, we may wish to add that only animals free of these bizarre behavior patterns should be considered adequately housed and/or reared.

In reviewing the breeding and rearing success of the great apes, it is encouraging to note that considerable progress has been made. However, it is misleading to examine the record of great ape husbandry without reference to the success of *individual* species. While chimpanzees have been bred for many years with reasonable success, orangutans and gorillas have had a troublesome record in this respect. The first captive orangutan birth was recorded in 1929, while gorillas did not reproduce in captivity until 1956 (cf. Bourne and Cohen, 1975). In all of the great apes, second generation births are the most difficult to achieve. That is, apes born in captivity do not readily reproduce. The reasons for this will become apparent as we discuss this issue further. It is with these captive-born animals that rearing problems are most apparent, for once the problem of breeding has been solved, the mother must then exhibit proper parental care.

Differential breeding success in respective great ape species can be attributed to a number of factors. For example, the success of the Cincinnati Zoo has been partially attributed to diet. It is also clear that there are differences in the amount of sexual behavior normally exhibited by chimpanzees, gorillas, and orangutans (cf. Nadler, 1977), and in relation to this, captivity may differentially affect reproductive behavior in these species (cf. Maple, 1975; Maple, 1977).

THE NATURE OF APES

To properly plan for ape-keeping, it is first necessary to understand the behavioral characteristics of these creatures, in captivity and in the wild. Regrettably, there are many gaps in the literature, and about an ape as rare as the pygmy chimpanzee, next to nothing has been written. Nonetheless, we have learned enough so that some useful generalizations can now be made.

The Chimpanzee (*Pan troglodytes*)

The most studied of all the great apes, in laboratory, zoo, and field, are the common chimpanzees. The temperament of these primates is at once their most endearing and annoying trait. Highly responsive to human interest, chimps rarely fail to entertain. In the zoo, they are well known for their propensity to shower human visitors with saliva, water, or excrement. While some may find this entertaining, it is neither a healthy nor a happy state of affairs. In glass-enclosed cages, chimps are prone to jump and kick against its surface, making a considerable noise. This activity is also a "delight to the public" as Mottershead (1959) has pointed out. However, in my opinion, one of the great dangers of this overzealous responsiveness is that it encourages the public to tease other, more sensitive animals. At the Atlanta Zoo, I have witnessed the face-making, lunging threats of chimpanzees toward visitors and the uproarious human laughter which follows the chimpanzees' response. Sure enough, as they make their rounds, other animals are similarly treated in hopes of a reaction. Many of the Atlanta macaques attack their cagemates at the sight of a crowd, and I am convinced that public teasing is at fault. However, even on islands such as those at England's Chester

Zoo (Mottershead, 1959) or The Kingdoms Three Animal Park in Atlanta (Maple et al., 1978), chimps persist in throwing debris. Such behavior is not out of character for chimpanzees, since in the wild they tear up and throw plants, rocks, sticks, and even available man-made objects during their aggressive displays (Goodall, 1965; see also Maple, 1974). If aggressiveness to the public is to be reduced, animals will need to be reared in enclosures where human harassment is at a minimum. Without the opportunity to aggress the public (and the reduced need to do so), chimpanzees are free to socialize among themselves.

In addition to their temperament, chimpanzees are characterized by a considerable degree of intelligence. Because the chimpanzee is highly manipulative both in captivity (cf. Köhler, 1927) and in the wild (Goodall, 1965) where they have been observed engaging in simple tool use, the captive habitat must be sufficiently complex to prevent boredom in these creatures.

A third feature of chimpanzee behavior is their highly social nature. To satisfy the needs of this highly emotional, highly intelligent, and highly social animal, it must be housed with companions. In particular, these apes should benefit from a rich social environment in their early years, hence mother rearing is extremely important.

The Gorilla (*Gorilla gorilla*)

Although research has been done on wild mountain gorillas (*G. g. berengei*) (cf. Schaller, 1963; Fossey, 1971), only a few are exhibited in captivity. Curiously, on display throughout the world there are about 400 lowland gorillas (*G. g. gorilla*) about which comparatively little is known in the wild state (but see Sabater Pi and Jones, 1967). All comparisions of captive with wild gorilla behavior are therefore confounded by potential species differences.

In captivity, the lowland gorilla is considerably less active than is the chimpanzee, is difficult to breed, and is especially prone to coprophagy. In the earlier years of their exhibition, gorillas often fed to excess, resulting in extreme obesity. The gorilla *Gargantua* attained the incredible weight of 550 pounds. The two mountain gorillas displayed at the San Diego Zoo both attained weights of over 600 pounds. Even when trim, gorillas are huge creatures, with males typically weighing from 350–400 pounds. Females are considerably smaller, generally 150–200 pounds. The muscular features and stern facade of the gorilla have given it a reputation for ferocity and great strength (Bourne and Cohen, 1975). Only the latter is true. In the wild and in captivity, gorillas are reasonably peaceful creatures. However, sexual encounters can be quite rough (cf. Hess, 1973; Nadler, 1977).

In the wild, mountain gorillas are found in "family" groups led by one dominant male and generally composed of one to six females with infants and other offspring. Gene flow is apparently enhanced by the transfer of some mature males and females who eventually leave or are expelled from their natal group by the resident silverback male. In the wild, adult males do not often interact with offspring. However, recent observations in captivity suggest that there is a considerable potential for intense social interactions between adult males and their offspring (cf. Hoff et al., 1977).

The Orangutan (*Pongo pygmaeus*)

The wild orangutan has been described as a solitary creature (cf. Rodman, 1973); however, such a description does not accurately portray the full social capacity of this ape. Clearly, the traveling habits of orangs are different from gorillas and chimpanzees. Adult orang males are generally found alone, and they maintain large territories which they defend from neighboring males. However, orangs have been observed during prolonged consortships with females (cf. MacKinnon, 1974). Thus, even in the wild, orangutans seem capable of strong social attachments. The social relations of wild orangutans have not yet been fully described, despite the notable and careful efforts of Rodman (1973), Horr (1975), MacKinnon (1974), and Galdikas-Brindamoor and Brindamoor (1975). As will become clear, the prevailing view of the orangutan as a lethargic, solitary, emotionless creature has become a self-fulfilling prophecy in captivity.

The orangutan is the most arboreal of the great apes, and, whether in captivity or in nature, it moves cautiously. The animal appears very deliberate in everything it does. Researchers who have attempted to assess the orang's intelligence find that temperament interferes with performance. However, with patience and hard work, it is possible to train orangs to emit an incredible array of behaviors, as can be seen in the Las Vegas circus act of Bobby Barosini who owns the only trained orangutan act in the world. The effects of circus life on behavior will be discussed in a later section of this chapter.

One can readily see that in comparing the behavior of the great apes, there are a number of differences in the respective species. The most active and responsive of the great apes is the chimpanzee; its active and responsive propensities present both benefits and problems. Between the orang and chimp in activity is the gorilla, which is quiet and easily bored. Special care must be taken to insure that gorillas are socially stimulated. Orangutans, the so-called solitary apes, are capable of much more than most people expect. To emerge from lethargy, orangs require stimulating social and physical surroundings. In many ways, as we will see, orangs and gorillas require many of the same arrangements.

HABITAT REQUIREMENTS

Spatial Requirements

In considering the spatial requirements of great apes, it should be stated at the outset that no captive habitat can approximate the spatious dimensions of their natural habitats. Clearly, the construction of such vast arenas would be impractical and prohibitively expensive. However, it is possible to provide environments which stimulate the activities that would occur in the wild.

Vertical Space. For all of the apes, an important habitat dimension is the vertical component. Orangutans are primarily arboreal, and they require elevated pathways in order to locomote in their characteristic fashion (brachiation). Habitats which do not permit brachiation, contribute to the lethargy which often characterizes this species in captivity. Field studies of chimpanzees (Reynolds, 1967; Goodall, 1965) have indicated that they are also inclined to an arboreal life, although less so than orangutans. While gorillas are predominantly terrestrial, they use trees on occasion, particularly when they are young. All of the great apes are known to construct sleeping nests, and they spend many hours forming these structures if they are given branches, hay, or some other form of browse. This activity is not only good for the animals, it is quite interesting for the viewing public. Platforms or raised sitting areas allow the animals to build elevated sleeping nests. The increased activity required for building elevated nests is obviously beneficial. By attending to vertical as well as horizontal space, the entire enclosure can be better utilized. A climbing apparatus and sitting perches increase the complexity and therefore the quality of the environment.

Cover. Where groups of animals are housed together (a preferred state of affairs), an increase in spatial volume can reduce crowding and subsequent social stress. Moreover, proper internal construction will take into account the need for *cover.* As Erwin et al. (1976) have pointed out, cover can effectively remove an animal from view, thereby providing a means to shorten conflict. Modified cement culverts, protruding walls, and room partitions are especially effective in providing refuge and privacy. Open sleeping dens have also been employed in this manner.

Flight Distance. Of particular importance in the construction of captive habitats is the requirement that the animals be given adequate spatial separation from human intrusion. While it is difficult to determine what an adequate distance may be, the best habitats provide for at least twenty feet of separation. Of course, an enclosure with adequate depth will allow the animal to establish its own minimum distance. As Hediger (1950) has written:

> Since flight reaction is the most significant behavior pattern of the wild animal's life in freedom, it must be a prime concern in captivity to give normal play to this vitally important reaction. This means giving the animal the chance to get away from man himself, at least to beyond its flight distance. The smallest cage in theory must thus be a circle of a diameter twice the flight distance.

ACTIVITY REQUIREMENTS

Necessity of Activity

In the wild, apes must travel great distances in order to acquire sufficient sustenance. Foraging for food is a kind of work in which all wild animals must engage. To encourage such exercise, it is necessary to devise ways of stimulating activity. Yerkes (1925) recognized the importance of this activity in his book *Almost Human*:

> Undoubtedly, kindness to captive primates demands ample provision for amusement and entertainment as well as for exercise. If the captive cannot be given opportunity to work for its living, it should at least have abundant chance to exercise its reactive ingenuity and love of playing with things. . . . The greatest possibility of improvement in our provisions for captive primates lies in the invention and installation of apparatus which can be used for play or work.

Captive apes need activity in order to prevent boredom and promote health. Without activity they will engage in unhealthy behaviors such as excessive self and social grooming, repetitive regurgitation of food, and coprophagy. The latter two behaviors are, in part, the result of reduced food intake which is necessary in order to prevent obesity. Were the animals active, restricted food intake would be unnecessary and coprophagy and regurgitation would be unlikely to develop.

Techniques for Inducing Activity

One of the promising solutions to such problems is the technology developed by Hal Markowitz (see Chapter 8 in this volume). These devices can induce locomotion, cooperation, and problem-solving efforts when designed properly. An interesting example of an especially creative (yet simple) device is the tug-of-war in operation at the Honolulu Zoo (see Fig. 9-2). With this device the gorilla is given exercise and social stimulation. Moreover, the public is directly exposed to the strength of a gorilla. Games such as this and those described by Markowitz allow public-animal interactions which are safe, indeed, beneficial for the animal. These innovative techniques clearly fulfill a need which has previously been satisfied through public feeding, petting zoos, and animal rides. For the animals that are provided manipulanda, it is advisable to change the game from time to time. In fact, where any objects are introduced for play, it is helpful to substitute *novel* objects periodically. As has been previously emphasized, a *changing* environment is to be preferred over a static one. An added benefit of many games is that the cognitive abilities

Fig. 9-2. Honolulu Zoo Director Jack Throp's drawing of a gorilla tug-of-war setup.

of the apes can be studied in the context of entertainment. Morris (1959) has pointed out the value of testing chimpanzee intelligence in view of the public, and Reynolds (1967) has further extended this view to include other forms of training:

> . . . young tractable apes can be used in training programs, circus-style, to brighten their lives and amuse the public. There is absolutely no reason why zoos should be shy of using circus techniques. These may result in healthier

animals and greater subsequent breeding success, and zoos should embrace them as a valuable management aid.

In view of these assertions, it is interesting to note that Paul Fritz (personal communication), after years of work with chimpanzee rehabilitation at the Arizona Primate Foundation, has concluded that former circus chimpanzees are more likely to breed than their zoo or lab counterparts. Regarding fatigue from overtraining, Morris (1959; see also Morris, 1964, 1969) has written:

> Far from being overworked, exhausted chimpanzees, these demonstration chimps are by far the healthiest, most intelligent, and most alert that I have ever seen in captivity. They obviously benefit tremendously from their varied and complicated activities and one is immediately struck by the need for introducing some similar kind of occupational therapy for adult chimpanzees and for other primates.

HABITAT TYPES

Throughout the world, a variety of habitats have been created for the purpose of ape-keeping. Although my personal observations of European zoos are limited, I am indebted to my former student Susan Wilson who recently visited many European ape facilities and shared her findings with me. It is a tribute to the importance of captive breeding programs that Ms. Wilson was awarded a Thomas J. Watson Fellowship in order to evaluate European efforts on behalf of the apes.

Laboratory Facilities

Standard laboratory squeeze cages are completely inadequate for the keeping of apes. For clinical treatment or for invasive research efforts they may be required, but their detrimental effect on the behavior of the animals is clear. In fact, in many cases, isolation of an animal which is in need of care can actually contribute to its demise from the combined effects of its ailment and stress. Consider Köhler's (1927) early recognition of the problem of isolation:

> It is hardly an exaggeration to say that a chimpanzee kept in solitude is not a real chimpanzee at all. That certain special characteristic qualities of this spe-

> cies of animal only appear when they are in a group, is simply because the behavior of his comrades constitutes for each individual the only adequate incentive for bringing about a great variety of essential forms of behaviour.

Clearly, the biggest problems with laboratory housing are the problems of physical size and social isolation. Moreover, Reynolds (1967) has differentiated between the "hygiene" and "natural" schools of ape-keeping, the former of which predominates among laboratory workers. The benefits of a hygienic habitat are obvious, but the stark cement and steel enclosures are generally not conducive to social activity. This is not to say that cement and steel per se are to blame. A spatious and creative cement and steel enclosure could be both hygienic and natural, but few laboratories have endeavored to build such facilities. It is encouraging to note, however, that agencies of the United States government recently agreed to develop a facility for chimpanzee rehabilitation and captive propagation. Some of the suggested designs for such facilities included innovative concepts such as large and well-equipped exercise areas adjacent to home cages. The laboratory worker must contend with the demands of ready access, restraint, and control of the subject and still meet its psychosocial and physical needs. A ready compromise has rarely been achieved, but it is an achievement which is highly desirable and not without promise of success.

Zoological Parks

The problems of the zoo habitat are quite similar to those of the laboratory. However, some of these common problems are even more difficult to manage in the zoo setting. For example, the health of captive apes is difficult to protect when the public is permitted proximity to them. The modern solution to this pressing problem has been the use of glass-fronted enclosures. When the surface is strong and clean, a satisfying view of the animals is created, and they are relatively well isolated from human disease. The glass front also serves to protect the public from flying fecal projectiles, spitting, and water throwing which are common features of captive ape behavior. A drawback of windows is that they seem to encourage the reduction of barrier distance so that the public may get a close-up view of the animals. Chimpanzees often respond to these public invasions by kicking and hitting the glass which, as we have seen, encourages public teasing. Glass-fronted enclosures therefore may inadvertently violate flight distance requirements.

Within the zoo enclosure, whatever barriers may be constructed, a variety of techniques may be used to enrich the surroundings. To simulate trees and natural cover, steel and concrete can be employed in creative ways. A good example of this is the hamadryas baboon enclosure at the Madrid Zoo. In this setting, the complexity and utility of the structures are striking. Equally functional is the large wooden apparatus which was recently erected at the San Diego Zoo for orangutans and the complicated, but smaller, version which has been standing at the Phoenix (Arizona) Zoo for several years (Fig. 9-3). Although subject to wear, wood lends a natural appearance to any enclosure. Even in the construction of human playgrounds, wood structures have become very popular. In the construction of wooden apparatus, volunteers from the community may be employed as has been done by Lee White at the San Francisco Zoo (Fig. 9-4). A useful manual for constructing such structures has been written by Hewes (1974). At the Zurich Zoo in Switzerland, bamboo has been used to construct arboreal pathways for a variety of species. Another way to use bamboo is as a background, where it can be separated from the animals by wire or glass. Although an illusion, the use of growing foliage as background creates a superior educational display, and the growth may be culled periodically to obtain browse.

A subtle dimension to captive habitat design is that of color. Whether a colorful background is important to the apes is unknown, but the public's perception of these surroundings can certainly be improved by attention to color. It should be admitted, however, that cosmetic improvements are no substitute for enlightened design. Furthermore, it is clear that aesthetic considerations can often obscure functional considerations. This is often the case when well-known designers are employed to create *works* of art, rather than *working* art (Sommer, 1974).

Safari Parks

The concept of outdoor, seminatural safari parks has flourished in the last decade. Because these parks allow the animals a considerable

Fig. 9-3. An example of the creative use of wood to produce a complicated arboreal habitat for orangutans at the Phoenix Zoo. (Photo: T. L. Maple.)

amount of space, there are a number of difficulties where apes are concerned. Because of the apes' arboreality, the fences must be equipped with sheet metal topping to prevent escape. Because of the great strength and manipulative abilities of the apes, the fence itself must be exceedingly well constructed. Due to the cost of such alterations, "new zoo" designs for apes have favored islands surrounded by water or dry moats (Table 9-1). Often, moated, outdoor exercise

Fig. 9-4. Community-built enriched habitat for chimpanzees at the San Francisco Zoo. (Photo: L. White.)

areas adjoin indoor housing, particularly in cities where the climate is not conducive to ape-keeping.

The more economical solution to ape-keeping is the use of natural or man-made islands, surrounded by water. In all cases, however, the island *must* be equipped with protective housing, to provide both cover and heat. Moreover, these areas must be secure in order to facilitate capture. This is best accomplished when the animals are locked up at least once each day for feeding and close-up inspection. While the rain and cold are an ever-present danger to health, the water barrier is equally hazardous. Opinions vary as to the reasons for drownings, but few apes survive a fall into water, since they are apparently not capable of swimming.

For five years (1972–1977), the Yerkes Regional Primate Research Center maintained chimpanzees on Ossabaw Island, off the coast of Georgia. Despite an optimistic report by Wilson and Elicker (1976), the animals were removed after a number of animals had died. The danger of drowning, and the difficulty of maintaining the animals un-

Table 9-1. Barrier Dimensions for Outdoor Enclosures for Great Apes (From Reuther, 1976)

	Zoo	Actual Depth[a] (cm)	Recommended Min. Depth (cm)	Actual Width (cm)	Recommended Min. Width (cm)	Comments
Orangutan *Pongo pygmaeus*	Dudley	170		460		wet moat and hot wire (Roots, 1963)
	N.Y. Bronx	270	270	430		water level 60–150 cm
	Toronto Metro	330	370	370	370	
	Phoenix	370	370	380	380	
	San Francisco	370		490		
	Twycross	370	240–270	370		animals have reached up to 210 cm
	Antwerp			400		wet moat; wiremesh barrier separates animals from deeper section (Van den bergh, 1960)
	Kansas City	430	370–430	430	430	
	Miami Goulds	460		460		water level 30 cm (DuMond, 1970)
	Baton Rouge			240–1520	240	wet moat surrounds an island
Chimpanzee *Pan troglodytes*	Chester	90	120–150	430–520	460	wet moat and hot wire (Mottershead, 1960)
	Arnhem	130		800		wet moat (Van Hooff, 1973)
	Dudley	170		460		wet moat and hot wire (Roots, 1963)
	Ibadan Univ.	260		490		water level 200 cm at

Table 9-1. (Continued)

	Zoo	Actual Depth[a] (cm)	Recommended Min. Depth (cm)	Actual Width (cm)	Recommended Min. Width (cm)	Comments
						public side + 60-cm dry wall above; hot wires at 105 cm depth (Golding, 1972)
	Berlin W.	300	300	360	360	
	Tulsa	300	300	610	460	wet moat (proposed)
	N.Y. Bronx	300		430		shallow wet moat (Crandall, 1964)
	Pretoria	320		160		hot wire in front
	San Francisco	370		490		
	Duisburg	410		440		
	Kansas City	430	370–430	430	430	
	Antwerp			400		wet moat (Van den bergh, 1960)
Gorilla *Gorilla gorilla*	Dudley	170		460		wet moat and hot wire (Roots, 1963)
	Frankfurt	180		500		wet moat led to accidents and has since been filled in and replaced by 400-cm high glass wall (Scherpner, 1967, 1971)
	Kyoto	210		500		wet moat with 10-V

					electrical potential (Watanabe, 1971)
Ibadan Univ.	260		490		water level 200 cm at public side + 60-cm dry wall above; hot wires at 105 cm depth (Golding, 1972)
N.Y. Bronx	270	270	430		water level 60–75 cm
San Diego WAP	300	300	300	300	
Munich	320	320			glass wall
Toronto Metro	320	490		300	
Phoenix	340	300	400	400	
Jersey	370	370	340		
Oklahoma	370	370			
San Francisco	370		460		
Twycross	370	240–270	370		animals have reached up to 210 cm
Houston	370	300–370	380–430	240–300	dry moat (Werler, 1975)
Kansas City	430	370–430	430	430	
Antwerp			400	360	wet moat (Van den bergh, 1960)
Chester			430–520	460	wet moat and hot wire

[a] Depth of moat/wall at outer perimeter.

der absentee management conditions, contributed to the decision to remove them. Obviously, it is better to use islands which can be easily managed on a daily basis. If a large natural island can be properly managed, there are a number of advantages, including the following ones mentioned by Wilson and Elicker:

> 1) Physical health can be easily monitored and any necessary medical treatment can be administered.
> 2) Intensive and extensive behavioral observations should be less difficult than in the wild state.
> 3) The maintenance cost per individual is less than in the captive state, as is the cost of building the initial facilities.
> 4) Reproductive rates are at least as high as those observed in captive settings.
> 5) The infants produced are psychologically and behaviorally more normal.
> 6) Previously experimental, behaviorally abnormal adults will experience some degree of rehabilitation towards the more normal species-typical behavior observed in wild populations.

The island concept has also been discussed by Reynolds (1967) in the advocacy of what he has called *apelands*. Like the commercially popular *marinelands*, these apelands, as Reynolds sees it, would provide for public exhibition and entertainment. When not in the public view, the animals would be allowed free access to a small forest with perimeter feeding stations where animals could be checked and/or captured. Daily observation by wardens would be a regular maintenance task, while lab work would be conducted in adjoining buildings. Those animals which were used in research would be periodically shuffled from forest to lab and back again. In suggesting the necessary elements for a successful enclosure, Reynolds mentions the following:

> . . . trees, a grassy paddock, moat with electric wire, cover (rocks) and dividers, rain shelters, fresh-cut branches for nest construction, interspecies housing (where space allows it), indoor housing with climbing materials, free access to indoor and outdoor facilities, ad lib food, and access to conspecifics.

The electric wire mentioned by Reynolds is one solution to the drowning danger posed by water. With this technique the animals are discouraged from approaching the water. At Kingdoms Three Animal Park in Atlanta, our research team constructed a similar device in order to discourage activity in the water (Fig. 9-5). Our barrier is a chain supported by pipes which rings the island, giving the

Fig. 9-5. Submerged chain barrier around chimpanzee island at Kingdoms Three Animal Park, near Atlanta, Georgia. (Photo: T. L. Maple.)

animal an opportunity to grab onto something should it slip into the water. We have had no incidents since the installation of the chain, despite the fact that two apes had drowned at the park in previous years. Van den Bergh (1959) also utilized a barrier system to prevent drownings as follows:

> We have safeguarded ourselves against a similar occurrence (drowning) by placing a wire mesh barrier between the shallow area of water near the enclosure and the deeper section of the moat. Since its installation, the barrier has twice saved two very lively young gorillas which fell into the water and were able to get out again on their own accord.

Another problem with islands is that unless they are quite large, the inhabitants will soon eat their way out of house and home. Therefore, it is often necessary to build artificial structures which provide shade and refuge should foliage be depleted.

Where the weather is suitable for island living (tropical or subtropical settings are to be preferred), their advantages are obvious. Islands can be an inexpensive, natural alternative to the elaborate structures which have become the norm in modern zoo design.

HABITAT FUNCTIONS

Besides the primary functions of feeding, care, and capture, the captive habitat should be constructed so that public viewing and daily observations are unobstructed. The former is complicated by some of the animals' needs which we have previously discussed. The latter can be facilitated by building into the enclosure observation areas, closed-circuit television access points, and/or one-way observation windows. A well-equipped research program is an essential feature of successful captive animal management. Clinical judgments can be enhanced when the veterinarian is supplied with daily behavioral data. With regular observation, problems can be objectively verified and valid solutions are more likely to be found. Behavioral research on the apes is a matter of critical importance, since we are obligated to breed these endangered animals in captivity. As we will see, an understanding of behavioral problems can improve the state of the apes and insure that their captive propagation will be successful.

BREEDING PROBLEMS

For chimpanzees, gorillas, and orangutans, captivity has not been conducive to reproduction. Gorillas have been the most difficult to breed and this may be related to their shy disposition (cf. Schaller, 1963) or to their relatively reduced sexuality (cf. Nadler, 1977). For all of the great apes, it has been implied that a lack of early social experience is responsible for their breeding problems. This argument stems in part from the fact that most apes were captured in the wild as infants and hand raised. Since the work of Harlow (1971) has so clearly demonstrated that social deprivation can effect sexual behavior, the argument has frequently been applied to problems of the apes. However, it should be noted that although Rogers and Davenport (1969) found chimpanzees to be greatly affected by isolation, they also found some potential for recovery:

> Distortion of the environment in infancy and early childhood, including social isolation and substitute maternal care, has a drastic effect on the sexual behavior of the chimpanzee. It appears that the critical variable is early social experience. Second, some chimpanzees reared in total isolation can, in contrast to similarly reared rhesus monkeys, recover from behavioral injury to the extent that they can copulate.

Rogers and Davenport suggested that the chimpanzee's less rigid and less stereotyped pattern of sexual behavior (relative to that of the rhesus) allowed recovery. Moreover, they found that the sexual behavior of isolation-reared chimps was *less* drastically affected than that of a human-reared animal. This is probably caused by the development of a sexual preference for human caretakers over conspecifics ("zoomorphism"; Hediger, 1950; see also Maple, 1977).

Curiously, early social deprivation appears to affect individual animals (and perhaps species) differently. Orangutans at the Yerkes Primate Center, all captured as infants, have continued to breed with great success. Yet, other captive orangs have often proved to be poor breeders. The thirteen-year-old orangutans in the Sacramento (California) Zoo exhibited little sexual behavior until, as if by magic, the female gave birth in 1977. Could this have been an example of recovery as Rogers and Davenport explained it? Where the male is sexually aggressive as in the orangutan (cf. Maple et al., in press; Nadler, 1977) the male is likely to be the most affected by deprivation. That is, unlike the gorilla and chimpanzee, where the female may become sexually proceptive and initiate mating, the orangutan female's relative passivity is unlikely to compensate for male reproductive inadequacy (but see Maple et al., in press). In the orang, scientists such as Perry (1976) have been especially concerned with the lack of second-generation births, offspring from zoo-born parents. Interestingly, the first known second-generation birth, at the National Zoo in 1977, was the offspring of two *mother-reared* animals; the female of the pair was raised at the Atlanta Zoological Park. From this case and others, it appears that mother rearing is of paramount importance in the sexual development of orangutans, if not all apes.

A recurring and pressing problem, therefore, is the adjustment of great apes to hand- or nursery-rearing procedures. Where such procedures are necessary, the institution runs the risk that the animal will develop abnormal behavior patterns due to sensory-social isolation and/or environmental restriction. Two cases in point are our young male orangutans Anak and Bulan. Both of these animals have been nursery reared due to abuse by their mother and are under study by our research team. In the case of Anak, a stereotypic rocking behavior developed which, combined with his relative isolation from conspecifics, interfered with his social interactions.

We believe that this behavior is caused by a lack of motion stimulation normally provided by the mother's movements. In such cases, we recommend that the institution construct a mechanical surrogate which is capable of movement or arrange for extensive daily periods in which the infant is allowed to cling to a busy caretaker (cf. Mason and Berkson, 1975).

We can happily report that Anak's behavior has been improved by successfully introducing him into a group consisting of a ten-year-old nulliparous female and her twenty-year-old mother, currently nursing a $1\frac{1}{2}$-year-old male offspring. It is especially interesting to note that the young female has directed considerable "maternal" care behaviors toward Anak. An added benefit of this introduction is that Kesa (the young female) is learning appropriate maternal behaviors which should insure that her first-born offspring will receive adequate care.

Bulan's case is of even greater interest to us, since he has taught us a great deal about both infants and mothers. Born in the winter of 1976, Bulan was the second infant to be abused by its mother, Lada (the first was Anak). The pattern of abuse included both peeling of the stomach skin (following umbilical chewing) and removal of the infant's fingernails by chewing.

We attempted to alleviate this problem by providing the mother with quantities of hay, in the hopes that she would pay less attention to her infant. This solution was satisfactory for several days, but as her interest in the hay diminished, her propensity to peel the infant increased. In the end, we were forced to remove the infant for repair. In our nursery, Bulan developed self-clasping behaviors, whereby he habitually grasped his own arms with his hands.

At the age of six months, Bulan presented us with the unique opportunity to attempt an adoption. The twenty-year-old female orang Sungei, residing at the Atlanta Zoo, had recently lost her one-year-old infant, and we quickly acted to introduce Bulan. We were especially gratified to learn that Sungei would so readily accept a new infant. She eagerly retrieved him, held and examined him. Although she appeared somewhat surprised at his helplessness (relative to her older infant), she adjusted to the differences with noticeable facility. Bulan, however, proved to be an inadequate subject for adoption, in that his self-clasping interfered with ventral clinging. Because of this propensity, Bulan was unable to cling to his new "mother." Moreover, he also exhibited a deficiency in locating the nipple, suckling only once during the eight hours with Sungei. Despite the excellent care provided by Sungei, Bulan's helplessness required that we return him to the Yerkes nursery.

What these two case studies tell us is that restricted rearing conditions can produce extremely resistant behavioral problems. We are encouraged, however, by the potential of an "aunting" strategy whereby adequate social environments can be reconstructed. The proper degree of nursery stimulation may well reduce the magnitude of the developmental obstacles which are described here.

Another possible problem in breeding captive apes is that animals which are reared together may become habituated to each other and fail to exhibit sexual interest. Most zoo workers state that apes readily form preferences and are often found to be incompatible with particular animals. A further dimension to this question was suggested in another context by the psychiatrist, Temerlin (1975), who raised a female chimpanzee in his home:

> I had thought that Lucy would be "imprinted" on me and direct her sexuality toward me when she reached sexual maturity. . . . But sexual imprinting did not occur in the predicted fashion, and the redirection of her sexuality toward men outside the family suggests an incest taboo, at least in female chimpanzees.

Could it be, therefore, that heterosexual pairs reared together from infancy develop a "brother-sister" relationship, which in some way inhibits mating? It may well be that in some instances this is exactly the case.

STIMULATION AND COMPANIONSHIP

If we can apply the findings of Harlow and his co-workers, peer experience may be as important to normal social development in apes as it is in monkeys. Therefore, when possible, peer interactions ought to be arranged, either through loan arrangements or by temporary interspecies mixing. In many nursery and laboratory settings, interspecific companionship has been successfully arranged, and is certainly to be recommended when mothers have rejected their infants (cf. Maple et al., 1978; Maple, 1974).

Stimulation from some source is to be recommended for all cap-

Fig. 9-6. Foliage serves as natural cover on the chimpanzee island at Kingdoms Three Animal Park. (Photo: T. L. Maple.)

tive animals, particularly young ones who must be hand or nursery reared. In some cases, television has been employed as a diversion, although its effectiveness has not been fully evaluated. In a similar vein, as one component in a multifaceted stimulation program, I collaborated with Sacramento Zoo Director William Meeker to project films of gorilla copulation onto the wall of the zoo's gorilla enclosure. The animals were aroused to activity by the events, but we could not be sure that it was effective in stimulating copulation. We only expected to stimulate greater activity per se, in the hope that increased activity would enhance social interaction. While no gorilla births resulted from this "experiment," shortly thereafter, the orangs residing in the next enclosure exhibited the remarkable recovery that I mentioned previously. Although I would not seriously suggest a connection, it is an amusing possibility.

CONCLUSIONS

Throughout our history of ape-keeping, humans have constructed occasionally good, but more often bad and even ugly habitats for these sensitive and intelligent creatures. In reviewing the literature, and the lore of ape-keeping, it is clear that we still do not have definitive, empirically derived evidence on the social effects of habitat. Still, there are some testable hypotheses, solid and compelling influences, and many useful observations on which to rely. There is much agreement on optimum habitats, and a general trend toward improvement in design, management, and display.

To aid further progress, we need more applied research on these questions and greater cooperation in bringing about these positive changes. Great ape breeding loans should be monitored by scientists

Fig. 9-7. Chimpanzees have not yet destroyed this natural wood shade structure in one year of use at Kingdoms Three Animal Park. (Photo: T. L. Maple.)

in order to define problems and procedures, and there is a pressing need to share information and encourage outside investigators from our universities.

While I have suggested a great deal here, I am painfully aware of the need for more convincing data, and ultimately more compelling arguments. Habitat *can* be assessed as an independent variable which affects behavior. We can count the number of corners, measure area and volume, ennumerate manipulanda, and record activity. We can conduct pretests and posttests, and we can combine our notes to increase the number of subjects and observations. We *can* do research

Fig. 9-8. Chain-link barrier allows caretaker to approach island without fear of harassment by ape inhabitants. (Photo: T. L. Maple.)

Fig. 9-9. If not prevented by barriers, apes develop begging behaviors. (Photo: T. L. Maple.)

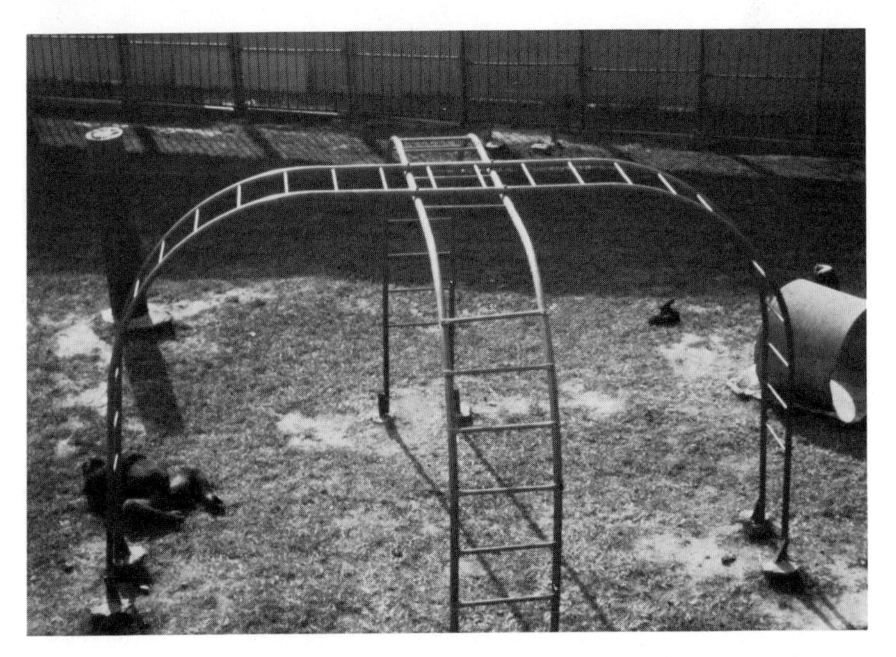

Fig. 9-10. Climbing structures for gorillas at the Yerkes Regional Primate Research Center Field Station, Lawrenceville, Georgia. (Photo: T. L. Maple.)

on the effects of captivity. Indeed, we *must* do it. As Yerkes and Yerkes (1929) expressed it:

> In the past, knowledge of anthropoid life has grown haltingly, irregularly, uncertainly, because of fragmentary, unverified, and often unverifiable observations. Because of adverse conditions, investigations have been relatively unsatisfactory: witness, attempted contributions to knowledge of courtship, mating, and other important aspects of the reproductive cycle, of life history, rate and conditions of growth, mode of life, heredity and acquired modes of response.... Obviously, many things have been done poorly, though at great pains, which under carefully planned and appropriate conditions might have been done well.

Now is the time to rectify this situation and to initiate systematic, cooperative inquiries into the state of the ape in captivity. The challenge is an urgent one if we are to successfully insure that these endangered anthropoids are to survive the onslaught of human civiliza-

Fig. 9-11. A dry-moated enclosure for orangutans at the Los Angeles Zoo, with no arboreal possibilities. (Photo: T. L. Maple.)

Fig. 9-12. An example of an inadequate arboreal habitat for orangutans. (Photo: T. L. Maple.)

tion and the cruel manifestations of our relentless history of material progress. As Sommer (1974) has concluded:

> If living creatures cannot be left in their original habitat, the least that can be done is to place them in natural and responsive surroundings–natural so that their character is not warped, and responsive so that their individuality and creativity are firmly respected. (p. 69)

REFERENCES

Berkson, G. Development of abnormal stereotyped motor behaviors. *Dev. Psychobiol.* **1**(2): 118–132 (1968).

Bourne, G. H. and Cohen, M. *Gentle Giants.* New York: Putnam, 1975.

Breland, K. and Breland, M. *Animal Behavior.* New York: Macmillan, 1966.

Ellenberger, H. F. The mental hospital and the zoological garden. In J. and B. Klaits (Eds.) *Animals and Man in Historical Perspective.* New York: Harper & Row, 1974, pp. 59–92.

Erwin, J., Mitchell, G., and Maple, T. Abnormal behavior in non-isolate reared rhesus monkeys. *Psychol. Rep.* **33**: 515–523 (1973).

Erwin, J., Anderson, B., Erwin, N., Lewis, L., and Flynn, D. Aggression in captive groups of pigtail monkeys: Effects of provision of cover. *Percept. Mot. Skills* **42**: 319–324 (1976).

Fossey, D. More years with mountain gorillas. *Nat. Geographic* **140**(4): 574–585 (1971).

Galdikas-Brindamoor, B. and Brindamoor, R. Orangutans, Indonesia's "people in the forest." *Nat. Geographic* **148**: 444–473 (1975).

Goodall, J. Chimpanzees of the Gombe Stream Reserve. In I. Devore (Ed.) *Primate Behavior.* New York: Holt, Rinehart and Winston, 1965, pp. 452–473.

Hancocks, D. *Animals and Architecture.* London: Hugh Evelyn, Ltd., 1971.

Harlow, H. F. *Learning to Love.* Chicago: Aldine, 1971.

Hediger, H. *Wild Animals in Captivity.* London: Butterworth, 1950.

Hess, J. P. Some observations on the sexual behavior of captive lowland gorillas. In R. P. Michael and J. H. Crook (Eds.) *Comparative Ecology and Behavior of Primates.* New York: Academic Press, 1973, pp. 507–520.

Hewes, J. J. *Build Your Own Playground.* Boston: Houghton Mifflin, 1974.

Hoff, M. P., Nadler, R. D., and Maple, T. The development of infant social play in a captive group of gorillas. Paper presented at the Meeting of the American Society of Primatologists, Seattle, Wash., 1977.

Horr, D. A. The Borneo orang-utan: Population structure and dynamics in relationship to ecology and reproductive strategy. In L. A. Rosenblum (Ed.) *Primate Behavior: Developments in Laboratory and Field Research*, Vol. 4. New York: Academic Press, 1975, pp. 307–323.

Köhler, W. *The Intelligence of Apes.* London: Metheun, 1927.

Lang, E. M. The birth of a gorilla at Basel Zoo. *Int. Zoo Yearbook* **1**: 3–7 (1959).

MacKinnon, J. The behavior and ecology of wild orang-utans (*Pongo pygmaeus*). *Anim. Behav.* **22**: 3–74 (1974).

Maple, T. Basic studies of interspecies attachment behavior. Ph.D. dissertation, University of California, Davis, 1974.

Maple, T. Fundamentals of animal social behavior. In E. S. E. Hafez (Ed.) *Behaviour of Domestic Animals*, Vol. 2, 3rd ed. London: Balliere-Tindall, 1975, pp. 171–181.

Maple, T. Unusual sexual behavior of nonhuman primates. In J. Money and H. Musaph (Eds.) *Handbook of Sexology.* Elsiver: North Holland Biomedical Press, 1977, pp. 1167–1186.

Maple, T., Zucker, E. L., and Dennon, M. B. Cyclic proceptivity in a captive female orang-utan. *Behavioral Processes*, in press.

Maple, T., Zucker, E. L., Hoff, M. P., and Wilson, M. E. Behavioral aspects of reproduction in the great apes. In *Proc. 1977 Meeting of the American Association of Zoological Parks and Aquaria.* Indiana: Hills-Riviana, 1978, pp. 194–200.

Maple, T., Wilson, M. E., Zucker, E. L., and Wilson, S. F. Notes on the development of a young orang-utan: The first six months. *Primates* 1978, **19**(3): 593–602.

Mason, W. A. and Berkson, G. Effects of maternal mobility on the development of rocking and other behaviors in rhesus monkeys: A study with artificial mothers. *Dev. Psychobiol.* **8**: 197–211 (1975).

Michelmore, A. P. G. (Ed.) *Zoo Design 2: Proceedings of the Second International Symposium on Zoo Design and Construction.* Paignton, Devon, England: Paignton Zoological and Botanical Gardens, Ltd., 1976.

Morris, D. The new chimpanzee den at London Zoo. *Int. Zoo Yearbook* **1**: 20–23 (1959).

Morris, D. The responses of animals to a restricted environment. *Symp. Zool. Soc. London* **13**: 99–118 (1964).

Morris, D. *The Human Zoo.* London: Jonathan-Cape, 1969.

Mottershead, G. S. Experiments with a chimpanzee colony at the Chester Zoo. *Int. Zoo Yearbook* **1**: 18–20 (1959).

Nadler, R. D. Chimpanzee sexual behavior in relation to the gorilla and the orang-utan. In G. H. Bourne (Ed.) *Progress in Ape Research.* New York: Academic Press, 1977, pp. 191–206.

Perry, J. Orang-utans in captivity. *Oryx* **13**(3): 262–264 (1976).

Reuther, R. Barrier dimensions for lions, tigers, bears, and great apes. *Int. Zoo Yearbook* **16**: 217–222 (1976).

Reynolds, V. *The Apes.* New York: Dutton, 1967.

Rodman, P. Population composition and adaptive organization among orang-utans of the Kutai Reserve. In R. P. Michael and J. H. Crook (Eds.) *Comparative Ecology and Behavior of Primates.* New York: Academic Press, 1973, pp. 171–209.

Rogers, C. and Davenport, R. K. Effects of restricted rearing on sexual behavior of chimpanzees. *Dev. Psychol.* **1**(3): 200–204 (1969).

Sabater Pi, S., and Jones, C. Notes on the distribution and ecology of the higher primates of Rio Muni, West Africa. *Tulane Studies in Zool.* **14**: 101–109 (1967).

Schaller, G. *The Mountain Gorilla: Ecology and Behavior.* Chicago: University of Chicago Press, 1963.

Sommer, R. *Tight Spaces.* Englewood Cliffs, N.J.: Prentice-Hall, 1974.

Temerlin, M. K. *Lucy: Growing up Human.* Palo Alto: Science and Behavior Books, 1975.

Thomas, W. D. Great ape houses and grottos at the Omaha Zoo. *Int. Zoo Yearbook* **9**: 62–63 (1972).

Van der Bergh, W. The new ape house at the Antwerp Zoo. *Int. Zoo Yearbook* **1**: 7–12 (1959).

Van Hooff, J. A. R. A. M. The Arnhem Zoo chimpanzee consortium: An attempt to create an ecologically and socially acceptable habitat. *Int. Zoo Yearbook* **13**: 195–203 (1973).

Wilson, M. L. and Elicker, J. G. Establishment, maintenance, and behavior of free-ranging chimpanzees on Ossabaw Island, Georgia, U.S.A. *Primates* **17**: 451–473 (1976).

Yerkes, R. M. *Almost Human.* New York: Century, 1925.

Yerkes, R. M. *Chimpanzees: A Laboratory Colony.* New Haven: Yale University Press, 1943.

Yerkes, R. M. and Yerkes, A. W. *The Great Apes.* New Haven: Yale University Press, 1929.

Author Index

Subject Index